DOWN ON THE

THE LAST DAIRY FARMS OF NORTH STONINGTON

DOWN ON THE FARM

THE LAST DAIRY FARMS OF NORTH STONINGTON

MARKHAM STARR

Printed by

The North Stonington Historical Society

Preserving the history and culture of the local community.

FOWLER ROAD PRESS
49 Fowler Road
North Stonington, CT 06359
Telephone: 860-535-4413

ISBN - 10 0-9821685-3-5
ISBN - 13 978-0-9821685-3-0

First Edition Published November, 2010
Book Design by FOWLER ROAD PRESS
All Photographs by MARKHAM STARR
markhamstarrphotography.com

Cover Photo: Neil Main and Tim Blake discuss chopping corn on the Palmer Farm
Frontispiece: From the cover of the *Western Farmers' Almanac*, 1849. Artist Unknown
Back Cover: Barn at Cool Breeze, Dairy Cow, Carrol Miner packing silage with his 1942 Cletrac

This book is dedicated with love to my wife and best friend Sue.
Without her, none of my projects would have been possible...

Foreword

North Stonington has long been an agricultural community. Settlers first began moving into the area in the late 1600's when the expanding population along the coastline drove them further into the landscape. While the town was named in 1724, an extension of Stonington, it took another eighty-three years before it was officially incorporated in 1807. One year later, tax statistics showed the town's population to be nearly 2500 people. In the more than two hundred years since, the town's rural character has been maintained as the population has only more than slightly doubled. At its incorporation, more than fifty percent of the town's fifty-four square miles of land were given over to cultivation and grazing, with sheep being the most populous animal in town. At that time, there were some 3,400 sheep, 445 oxen or bulls, and nearly 1330 cattle in town.

During the course of the 1800's the town expanded with new industries. Fulling, carding, grist, and saw mills sprang up along with other businesses used to support them and their workers – stores, ironworks and the like. While living up to its name with regards to the production of field stones, North Stonington also allowed local families to prosper on its lands. Farms sprouted along its miles of roads, and even families who didn't farm for a living kept some animals for the food they produced. Chickens for the eggs, pigs for the meat, and often a cow or two for the milk they could provide. As with all New England towns, each family strove for self-sufficiency, eliminating the need, where possible, of having to purchase food and materials from others. Today, of course, life is much removed from what it was even fifty years ago. The age of the generalist has gone by, and we often rely on the work of others to provide us with life's necessities. Once local, the community in which we now live is global in nature. The time of self-sufficiency has long since passed.

Plate 1: First cut of spring hay on Cool Breeze Farm. This hay is square-baled and stored dry in a barn.

While North Stonington prospered and grew for a large part of the 19th century, gaining the title "Milltown" as a reflection of the many industries built here, its fortunes and population began to dwindle with both the rise of the industrial revolution and further expansion by settlers to the west. Workers and factories moved to bigger towns with more powerful sources of water and local farmers moved to larger plots of open land in places such as New York and Ohio. The town's population dropped by nearly half, and land reverted to forests once again. The countless farms along the town's roadways began to disappear, and the village's new role as a bedroom community expanded by the end of the twentieth century.

Many family-owned dairy farms were able to survive throughout the years despite the major changes affecting the town. As recently as twenty or thirty years ago, dozens of farms in town were still involved in dairy production. Milk was produced not only for home and local use, but also for consumption in nearby cities. As with all of New England, however, family owned dairy farms have rapidly faded from our landscape. North Stonington's many dairy farms have dwindled to the four generational farms pictured within this book, with the hopeful addition last year of a small start-up farm that began its milking operation with seven cows. All dairymen in New England face enormous economic pressures from the vast dairy producers of the West. Competing with farms with 100,000 head of cattle is a difficult business, but also a testament to the skills and abilities of the farmers that have survived. Beyond the obvious economies of scale that benefit the Western farms, however, is the fact that dairy farming is not entirely a free market endeavor. Dairymen face some governmental regulation on what they can charge for their products. The price of milk is largely set by the economies of the West, where the scale is large and feed grain is cheap. New England farms, forced to compete against a few companies controlling more than ninety percent of all milk produced, have quite understandably fared poorly.

Given the harsh economic realities our town's farmers face, I thought it important to document their way of life lest it too vanishes forever from our sight. The intent of this book is to provide a window into daily life on these farms. To that end, I have spent the past year wandering around their farms in an attempt to capture a little of what it takes to operate a commercial dairy farm in New England today. What I found was much more than perhaps I had initially expected. What I came away with was greater than the photographs and interviews within this book. Most importantly, for me, was a real appreciation for what they do on a daily basis – three hundred and sixty-five days a year. While I can only hope to capture a fraction of their lives in so short a work as this, I hope the reader gains at least a small appreciation for what they do in this extended portrait.

Plate 2: End wall an of open shed at Beriah Lewis Farm with a large corn silage pile off to the right.

Plate 3: Snow falling on fields at the Miner Farm on Chester Main Road.

The Farms

There are currently four family-owned and operated commercial dairy farms left in the town of North Stonington. Near the completion of this book, however, an additional dairy farm began operation despite the difficult economic situation now confronting all dairymen in Connecticut. This book's primary focus remains the four farms with generations of family history on the soil they now inhabit. While the oldest farm, begun in 1791, now boasts its eighth generation working with animals, the youngest two of these four farms still represent nearly a hundred years of toil in the same location. Dairy farming, it seems, is passed down to the next generation through genes as much as any other inherited trait.

This book is the result of following each farm's activities over the course of a year. Farming in New England remains cyclical in nature. The earth is prepared and crops are both planted and harvested before the land is left to sleep once again through the grey New England winter. Calves are born, raised for several years until they mature enough to join the milking herd, and eventually sold at the end of their productive years. While farmers perform the same tasks each year at roughly the same time, they are never certain what surprises the months ahead may bring. Their view of the farm's prospects must necessarily encompass the long run. The dairy industry's health cycles up and down as much as the hills rolling through the land they work. Boom years are always offset with busts. While some factors influencing their business are obvious to outsiders, such as weather or the health of their herds, many others are never considered by the general public. Take, for example, the detrimental effect the planting of ethanol corn in America's heartland has had on the world stage.

Despite the fact that ethanol takes nearly as much energy to produce as it releases, this year's corn production is poised to set a new record in the United States because of fuel subsidies available from government sources. While this rush to create ethanol to power cars does little to help with this country's energy needs, it has wide-ranging effects both on our economy and the world's food supply. Rather than producing grain for consumption by citizens and animals, farmers in the Midwest produced a new type of corn for fuel. Switching crops led to higher grain prices throughout the world and significantly increased the cost to produce a gallon of milk for New England dairymen. A large part of the expense in New England's milk production are the grains that must be purchased and the transportation cost of getting these essentials to the area from the West by truck. Farmers were heavily impacted by this change in planting – higher grain pricing and a rise in shipping costs for that grain. While these costs rose throughout the country, the price paid for a gallon of milk at the farm remained flat.

The town's farmers have certainly seen changes over the course of their many years in operation. Early ancestors oversaw the transformation of dense woodlands into open fields. The type and number of animals they raised varied over the course of time, as did the value of the products created with them. Fields once plowed with horses are now cultivated by Deere. The explosion in genetic engineering in the last twenty-five years has changed not only the way dairy cows look today, but also what they consume and the quantity of milk they produce. Raw milk is no longer set in cold-water springs to cool before shipping by horse and wagon to the local trolley line for transportation to Providence. And yet, with all of the significant changes to which each generation had to

Plate 4: Tractor and feed mixer wagon near the barn at the Palmer Farm.

adapt, the heart of farming remained much the same. Modern dairy farmers are as committed to their work as their ancestors of a century ago. They get up and go to work on their farms every day of the year. Stewardship of the land, animal husbandry, and the creation of an essential food for the rest of the community still shapes the core of their identities.

As any casual visitor to these farms will quickly discover, the motivating force that drives farmers from bed at four AM and keeps them going until seven, eight, or nine o'clock at night is certainly not money. As each farmer related to me over the course of the year, there are many easier ways to make a dollar in this world. What time and technology haven't changed over the course of many years is the love they have for the job they do. As with workers everywhere, they may jokingly complain of the tasks before them, but they have all remained in the business because they can't imagine anything else they would rather do. It is a passion that literally flows through them with their blood, and without it, they would soon leave the farm for greener pastures. It is only necessary to follow them as they go about their daily chores for proof that this is so.

The four farms left in town belong to the Palmer family, the Lewis family and two related Miner families. The Beriah Lewis Farm has the honor of being the oldest farm in town. Now in its eighth generation, the farm was purchased in 1791. The second oldest, Cool Breeze Farm, was bought by Niles Miner's great-grandfather in 1839. The Palmer Farm in the Clark's Falls section of town and the second Miner family farm on the top of Chester Main Road were purchased in the late teens of the last century, although both families have been farming for over a hundred years. These are the last of the heritage farms in town – all others have vanished for one reason or another. Carrol Miner recalls some of the farms that once surrounded him, now gone:

Used to be a lot of dairy farms. Wilkenson down there used to make milk when they was livin' there. Down where Ken Lattimore was, Mocha Herbert always made milk down there. Fred York down there was another one, down there where the barn is across the road, goin' up Hangman Hill. Then you head down where First lived, Partlow – they used to make milk, course it was in cans. Stella Stewart, goin' up Reutemann Road. Then you got Henry Madison, he used to make a little milk. Then you got Charlie Hillard, down there where Rough Ruddies is. Then you had Oscar Burgess, that's just the next farm goin' over the road there. Then you got out on the corner – Norberg – they used to have chickens there. Then goin' back to the village there, they used to have Carl Gere. That was about it if you went down that road. Yeah, there was a lot of farms around. You'd have a few chickens and one, two cows, so they didn't have to buy everything from the store. Cripes, we used to have chickens running right around the village down there. You had to watch out so not to run over 'em when cars went down there.

The fact that these four farms have weathered all of the storms they faced is a

Plate 5: A cow rests in the tall grass on a warm spring morning at Cool Breeze Farm.

testament to their owner's skill and dedication. Below is a general listing of the people who own and operate each of the four farms detailed within this book. This list only covers the principals from each farm and some of those who appear in the photographs – not necessarily all of the workers employed. It is included to provide a better understanding of the voices quoted within the book and of the people who appear in the photographs.

The Beriah Lewis Farm is now owned and operated by two brothers – Ted and Ledyard Lewis, along with their mother, Rosalind. Other family members actively working on the farm include Noah Lewis – Ted's younger son – and Hummer Lewis' daughters, Elizabeth and Catherine Lewis. (Hummer Lewis, brother to Ted and Ledyard, was an integral part of the farming operation until his death in 2005.) John Clegg is in charge of feeding and oversees the production of the corn crop, and Butch Reynolds, another long-time employee now retired, was in charge of the grass crops and maintenance. The Beriah Lewis Farm also hires many other farmhands throughout the year.

The Palmer Farm in Clark's Falls is owned by another set of brothers – George and John Palmer. George's son Asa also works on the farm when off from school. As with the Beriah Lewis Farm, they also employ several key people who oversee vital aspects of the operation. Tim Blake feeds cattle and does much of the fieldwork, while Mark Taylor is in charge of cows from birth through eight months of age. As with everyone working a farm, each person may do ten other things each day beyond these simple descriptions. As the second largest farm in town, John and George hire many other workers throughout the year to help with various tasks.

Cool Breeze Farm is entirely operated by four family members – Niles and Esther Miner and their two daughters, Patricia and Linda. Patricia is married to John Palmer, the younger brother of the Clark's Falls farm. As with all of the farms, they occasionally hire people to fill in on a one- or two-day basis when timing is critical or outside expertise is required.

Carrol Miner and his wife Betty operate the farm on the top of Chester Main Road, along with Carrol's nephew, Robert Miner, and Carrol's son Orrin. It is not uncommon to find people from one farm working on another, as they often trade time and labor with each other during busy times or in times of crisis. This is most common while harvesting, when more hands are required to get the crops cut and stored before bad weather ruins them, but each farm helps the other when needed.

Cutting corn on the Palmer Farm.

Grain storage silos on the Beriah Lewis Farm.

Corn and barns at the George & Carrol Miner Farm.

Cool Breeze Farm

Cool Breeze Farm, located on Hangman Hill Road, is certainly one of the most picturesque dairy farms in Connecticut. It is the second oldest, continuously operated farm in town, and is owned and worked by Niles and Esther Miner and their two daughters, Patricia and Linda. The farm is now in its fifth generation. As Niles says:

My great grandfather bought this farm in 1839. He was from Stonington. The marriage certificate – just a little slip of paper – is upstairs with my great grandmother's picture . She was from North Stonington. When he bought the place, the deed called for two houses. I suppose the other one was the cellar hole in our pasture. I never remember any house there. There used to be a lot of houses up through the woods. They all moved to town, I guess. I suppose they moved in for the mills and for work. 'Course down on that place, where Patricia lives, there was a mill pond – used to be a grist mill and saw mill there – so they said. There was a house just the other side of the brook up in there. The foundation and well are still there.

Niles was born, raised, and has spent his life on Cool Breeze Farm, and as with most children, fit right into daily life on the farm:

Plate 6: View across the upper fields towards Cool Breeze Farm on Hangman Hill Road.

As a child I used to be out in the barn with my father quite a lot. I had my cup, and before I was big enough to step across the gutter, I'd just go in and crawl through the stanchions to milk a little milk in my cup to drink. When I got a little older and big enough to start doing something, I had to go to school. They closed the school up by my cousins, George and Carrol, the year I started, so they transported me down to Center School, right alongside of the town garage. That's where I went to school.

As with most farms in town, dairy was often just one component of the family's livelihood. They may have had a handful of cows for dairy, a few beef cattle, sheep for the production of wool, and crops planted for consumption by the family or for sale to others. According to Niles, Cool Breeze Farm was no exception:

Plate 7: Niles Miner on one of the many tractors needed to run the farm.

There used to be a lot of apple orchards here, so they say. Right back of the barn, where the manure pit is, was always called The Orchard. This lot down the road is called the Old Orchard. Then we have a Lower Orchard, Middle Orchard, and Upper Orchard. There were apple trees all over. 'Course there

used to be a cider mill out next to the barn. I remember the cider mill, but I don't remember it running. The '38 hurricane took that down.

The early dairy business was, of course, run much differently from the way it is handled today. While today's focus is aimed squarely at milk production, things were much different for earlier generations:

I know my father told about when he and his brothers went to the schoolhouse right up here on the hill. They used to keep some cows up there – dry stock and heifers I suppose – and when they went to school they would let 'em out in the mornin' so the cows could go out and get a drink of water and so forth. Then, on their way home from school at night, they'd go up and clean the barn, put 'em in, and give 'em a few fork-fulls of hay I suppose. In the summer, the cows would go out and get their own feed. That's the way they used to do it. Of course they'd put up hay, and well, silage, 'cause they used to have that old silo in the corner of the barn.

The dairy business today is almost entirely centered on milk production, and

Plate 8: Hay on the second floor of the barn built in 1917 by Niles' father.

Plate 9: The cornfield below the barn at Patricia and John Palmer's house.

the care and feeding of the animals occupies the majority of their time. Milking cows are no longer left to forage for themselves in fields during the warmer months of the year. Instead, they have food mixtures designed by animal nutritionists filling their troughs around the clock. All of today's fields are planted with the herd's nutrition in mind. Modern technologies and innovations in such things as plant and animal genetics, planting and harvesting equipment, refrigeration, and even barn design have markedly changed life for dairymen. Niles and his cousin Carrol remember the days before electric lines connected them to the outside world:

I heard 'em tell about them having an old lighting plant that would give them power. It was in the cellar out in the barn. They'd run it and charge up batteries.

Niles' father was one of four brothers. The dairy portion of the farming operation got larger before three brothers left the homestead and began farming in new locations throughout town:

My father had three brothers. I guess they started dairy farming before the brothers left. I know before Herbert left – he's the one that bought the place where Patricia lives – they had two stables in the barn, one on either side. When Herbert left, he took one stable – what was in it I don't know – and my father kept the other one.

Major changes in the town's dairy farming started when powered machinery began to replace horses as the main locomotive source. As is still the practice among the four farms in town today, cooperation and the sharing of machinery, time, and talent were prevalent:

My father used horses. He had one team here and my uncle Herbert had one. My father and Herbert were quite close and they worked together. They used to go around filling silos – I guess all four of the brothers did. My father used to have an old Paypec cutter – 13 inches he called it. I guess that was the length of the knives. It used to run on one of those old one-lunger engines – a Foose. The cutter would blow the silage right up into the silo through a pipe. It would chop up the corn, and had panels on the fly-wheel which would blow it up in. In 1924 they bought a 1020 International tractor with steel wheels. The cutter with the gas engine would plug up if a big bunch of corn was fed into it, but once they got the tractor with a governor on it, it would slow down a little, but the governor would open up and keep it going. Chester Maine told them, "There boys – you got your power – now go get yourselves a cutter!" That's when they got a 16 inch one. From that they went to a chopper. Twas about the same thing except it was on wheels and you took it out in the field. It had two heads – a corn head to cut the corn and a grass head to pick up windrows of grass. Got the chopper in about '54 I think – maybe '53. You would take the chopped corn and dump it into a blower that would blow it up and into the silo. It had an auger that would take the silage in – it was kind of a dangerous machine. I know my uncle got his foot caught in it once. I think they were down to Lewises. Dave was quick enough to shut it off in time so his foot didn't go into the blower, but I guess it twisted it quite a little and laid him up a while.

As new equipment was purchased, the work load on the surrounding farms changed as each man brought his new machine to his neighbor's aid. Money was not generally exchanged for this help, but traded for labor in the busy season. Most of this traded labor involved getting the crops planted, harvested, or stored for the winter. Common tasks in the daily operation of the farm were generally handled alone, except in cases where health problems or natural disasters such as hurricanes put farmers in need of extra hands:

There were four of us that worked together – Ken, Llewellyn, Ray Hill, and myself. I had the baler and the chopper, and I used to cut the corn and bale the hay. We kind of exchanged work if anything. I don't remember much money changing hands. The trouble was, if the weather was good, everybody would mow and we'd have a lot of hay to put up. We didn't chop too much grass. Some people would put up grass silage. I guess George and Carrol and Morris up on the hill would put up some grass silage. They had a Paypec chopper and I think George had a direct-cut head for it. They could mow it and take it right in. Otherwise, handling the green stuff was quite a lot of work.

Esther Miner was born and raised just over the town's border in Voluntown. Although she had an aunt and uncle who farmed, her father didn't start out in the farming business:

My father, before farming, did mainly carpentry. He worked at a lot of boat yards and he worked at Connecticut Cabinet at the end. After all of us six kids got through school, my father went farming. He said until then, without a steady paycheck coming in, it was a little iffy when you had six kids to feed.

While they were growing up, Esther and her brothers, sisters, and mother helped prepare their land for the future:

We didn't have that big of a farm, but we started clearing the land and making fields, and making more and more, so when he went farming he had the land to farm on. The farm was in Voluntown – just the second house into Voluntown on Route 49. First house was my grandmother's, where he was born, and the second house was his. That was originally part of the farm, but when he got married that was his share. They gave him a piece of actually no-good land, but he made it good. My mother always wanted to go into farming. They didn't have a big farm. I think they milked, how many cows? Fifteen, eighteen, twenty – enough to make a living then. We had most of the land cleared before he went farming. We did that with the idea that they were going to go farming eventually, and that kept us out of trouble – kept us working. Kept us picking stones and cutting brush.

Esther recalls how she and Niles met:

My father was down here filling silo – helping. Niles cut corn for Ken and my father and Ray Hill, and they all worked together on all four farms. My father worked down here cutting corn and my mother worked down here tending to his mother. I was in California at that time. I don't know, how did we meet? I

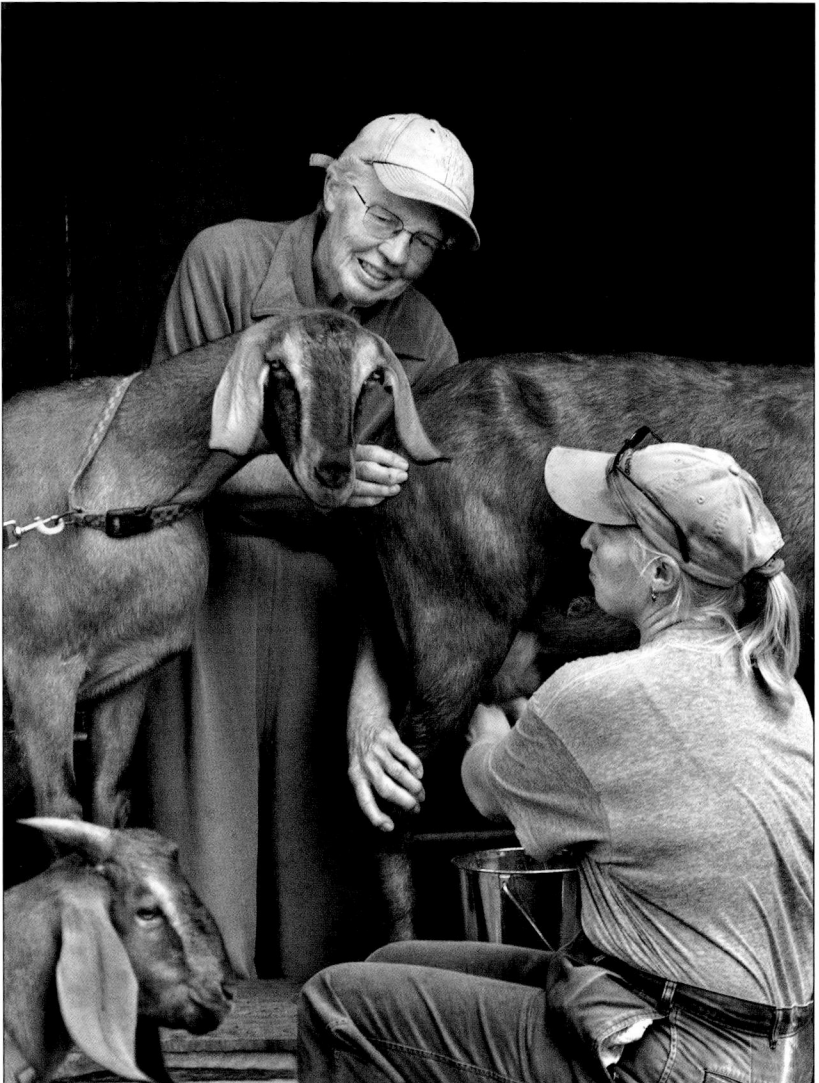

Plate 10: Esther and Linda Miner with a few of their goats.

know – I know! You were cutting corn for Franklin – his barn burnt down, and grandpa was going to drive truck, and you needed another truck driver – that was me. I came with my father. I was working for my father on his farm then. We got married in '69 – a looong time ago!

Known for her tremendous work ethic throughout the local dairy community, Esther had little time to simply ease into her life on the new farm:

I had to do everything, and I still do. A week after we were married he stuck me in the parlor – I never milked a cow in my life. I didn't milk cows at home. He'd disappear and then I'd have all these cows in here and he'd say, "Go to it, old girl!" And I milked all of them for seventeen years.

Esther was, at that time, a real pioneer in the dairy industry. It was still largely a man's world, and women were not often found doing the kind of work she did:

I think I was the first one around here – a woman – that would work milking cows. There were no women milking when I was milking. I was doing work on the silage

Plate 11: Esther is well known for the beautiful flower gardens she plants on the farm.

and I mean you pick stones! We got a stone picker so that he could go through and pick most of the stones. Of course the big ones you have to go through first with a bucket and get them off, and then the stone picker would pick most of them, not all of them. Of course it would push a lot of them in the ground, and then you'd have to harrow it again.

And every field that we made we had to make a path to, of course, so all of these paths that you walk on are all stone-based – dumped load after load after load after load and you'd make a path. And we'd get enough stones to make the path up to the field and beyond. And the next time we'd make another field – we would add on. So we made this path all the way up – a half a mile up. That's all stone-based – stones that we picked. Not only with the picker but by hand. I can remember we made a field over

trucks. I did the fieldwork – haying – picking up bales. We used to have so many stones to pick because we harrowed all the fields and we would pick stones for at least two weeks. Just picking stones, picking stones, picking stones.

Cool Breeze farm looked much different at the time of their marriage than it does today. Esther recalls the backbreaking work they did to make the farm into what we see now:

We'd keep making new fields. When we got married there was only about three or four fields here at the most. And we've cleared up all the other land since we've been married. When my father retired from farming, he was down here every day helping us, and mother would stay in and tend to Linda so that I could be out working. And it seemed like every year, for a long time, we made a new field. And whenever you make a new field around here you pick stones,

here. Stony? Oh my lord! And to this day that man still shakes his head 'cause he walked over there and said, "I don't know how anybody could be so stupid to make a field with that many stones!" And now it's just like all the other fields.

We've got the stones all picked, and we don't harrow it, because if we do we'll be picking stones again. All the walls that's got the big stones in them we built. We put a wall around all the fields that we made. We'd make our walls around with the big stones. We picked stones – let me tell you – we picked stones. And we'd pick all day long. We'd pick from chores in the morning till noon, eat dinner – and of course I'd have to get dinner while they ate, and then go back and pick until chore time at night.

We used to rent a lot of land – never had any open fields here – just a couple. Out where this pasture is here, you could walk from one end to the other and never put your feet on the ground – you'd just go from stone to stone to stone to stone. And there was so many brush and briar in there that the cows would just have these little paths. This field across the road – you couldn't take two steps in there. It was just briars and elderberry brush and you name it. And out where the calves are, that was where I had my sheep, and it was all those multi-flower roses and they'd have little tunnels underneath. Boy it was briery. This was a briar heap when we

Plate 12: The tank used to water the heifers is hooked to the tractor.

started in. Then he made that pond. It was just briars – you couldn't even get in there, and now look at it – the nicest pond in North Stonington. That thing doesn't freeze – there's a spring in there. It was open this winter all the time. Now I have got to go do chores… .

Raising a family, milking cows, driving truck, harrowing fields or picking stones, Esther has always been a force with which to reckon, and at nearly eighty years in age is never seen sitting still on the farm. Although they now have almost all the fields they need surrounding the farmhouse, Linda notes that they do rent one field:

We have a piece up on Pendleton Hill that we rent – we plant corn there. We try to stay at home, but I think we'd be short of feed if we didn't have that place. It's hard once you start running equipment up and down the roads. We used to use my aunt and uncle's place at one point but it was too much for us. If we were up there haying, the hay was going by here. If we were here, it was going by up there. I think we should concentrate right here. We usually plant about sixty acres of corn and fifty acres of hay.

As on all farms, size and age seldom exempt a family member from helping with daily life. Patricia helped on the new farm as a child, and Linda's birth simply added new twists to the daily routine, as Esther describes:

Patricia loved animals from day one. She always used to get up before school and go out and tend to the calves. She was up many mornings before I even got up. She'd go out and do the calves and she'd get back in so that I could go out and milk because Linda was a baby and somebody would have to be around. And then we'd put an inter-com from the bedroom to the milk parlor – Linda's bedroom – so I could even hear her breathe. That's why I was always on such a strict schedule. I had to start at a certain time so I could get through at a certain time so I could get back in before Patricia had to get on the bus. And when she kinda got bigger, I took her out in a carriage and she'd sleep while Patricia was playing sports at night. She was on a softball and a basketball team. And when she got bigger I put her in the stroller and she was out there, so I not only milked, I babysat. We had about twenty five to thirty cows. About twenty-five most of the time because I figured I wanted to get it done in an hour, from the time I started until I got through. Niles cleaned up – I didn't have time to finish cleaning up. I did the milking and I'd run back and make Patricia's lunch and get her breakfast so she could get to school.

Patricia was thirteen when Esther and Niles married, but not new to farming:

I didn't grow up on this farm, so when I was a real little kid, the farm I remember was my grandfather's farm up in Voluntown. My mom would work up there with her father

Plate 13: Niles chopping hay early in the summer.

– my grandfather. His parent's farm is the big white farm just as you get into Voluntown. You go up Rt. 49 – you go past the big yellow farmhouse. The next farm, the big white farmhouse – that was my great-grandparents'. They built that house – Birdsey Palmer Sr., and when my grandfather got married, he had property next door, and he had a dairy farm there. It's now a horse farm, just this side of where Sand Hill turns off. It was always neat – I really liked it. So when my mother married Niles, I thought it was the greatest thing that could ever happen because we ended up on a dairy farm.

Along with the normal activities in which school-age children are engaged, Patricia took care of the calves when at home:

When we moved on the farm I got to do the calves – that was pretty much my chore. My mom would milk, so I would be out in the morning before school tending to the little calves, and then when I got home in the afternoon, tend the calves again, and then help out with whatever had to be done. In the summertime, help with the haying, picking rocks – which was absolute torture – I hated

that! It's about the only thing I didn't like about farming when I was growing up. I didn't do any milking at home – my mom did the milking. I milked on other farms after I got out of college and I went to New York state for a while. When I moved back here I started milking here as well.

As Linda grew old enough, she too began to help out with the chores, although one impediment threatened to keep her from following in the family footsteps:

I remember unloading hay. If you were bigger than the hay bale you unloaded hay. I would be out on the wagon, pushing them down to whoever was putting them on the conveyor. When I was a kid I had terrible allergies to the cows, so I tended not to go near them. Just getting near a cow, my eyes would water and I'd start sneezing. I seem to have grown out of it now.

As she mentioned, she was always on the farm and there was always work to do, but:

It was kind of neat because my parents always worked, but they were always home too. My father did the fieldwork, and my mother always milked the cows. They worked a lot with Ken Lattimore who lived next door at the time, so they farmed together. He had his own herd, but they did the fieldwork – the baling and the cutting of the corn together. My father had the equipment and Ken would always help with the haying, and when they were haying down there we'd go down and help unload it.

Plate 14: Calves at Cool Breeze Farm.

In college, Linda majored in human development and family relationships and entertained thoughts of not going into the family business. She and her sister, however, eventually returned to the farm to help her parents on a full-time basis. While every farmer on every farm must be capable of performing dozens of tasks daily, there are two general categories of work into which most seem to fall in their life on the farm. There are crop people and there are cow people. As nearly everyone working on the town's farms commented at one point or another, you hardly ever find both in the same person. For this reason, most farms break down the long list of responsibilities along these two major lines, although there is always some overlap and additional work. As Linda notes:

We kind of do whatever needs to be done. My sister handles the nutrition – the feeding of the cows – and she keeps the cow's records, and I work on the books and paying the bills and doing the taxes and all that fun stuff.

Patricia also has her preferences for the workday:

I try not to get too involved in crops. I just like working with the cows. Niles – he's the crop man. There will be books just

Plate 16: Water and open farmland are important to other species in the area, such as this otter.

like the bull books on seed corn. He'll sit there and analyze all these different traits. There's so much knowledge out there genetics-wise now, whether it's on plants or on cattle. There have just been tremendous changes over the last twenty years in the way crops are grown. That's pretty much his area of expertise – he studies the corn and makes a choice. You've got hundreds of varieties, different companies to chose from, so he'll pick and choose there – the variety of corn that he wants to put in. To some extent on the grass too – you've got different choices of different grasses. He pretty much does that. I don't want to sit down and study a corn silage book. It just doesn't interest me at all. He's not big on the cows – he's more on the equipment and the fieldwork. I'd just as soon not do any fieldwork – I do what I have to do. If he needs somebody to harrow, or somebody to pick stones or rake hay we all pitch in and do it. He's done all the planting. Last year I did spray a little just because he wasn't here and it had to be done. I'm not crazy about doing that, but he wouldn't want to have to go into the parlor and milk the cows. That's not his area of interest. It works out – if you've got three or four people that all want to do the same thing and no one wants to go and do the other, then you're kind of hurting. So it works out. Fieldwork is his cup of tea and the cows aren't. Well that's OK, because my mother is real good with cows, and I'd rather work with cows, and Linda is really good working around the cows, so we don't need him in the parlor. You kind of have to be a jack of all trades, but it's enough for me to just concern myself with the cows. I don't need to make all the decisions on the crop. He can focus on that and really do a good job and he doesn't have to focus on, "Are we breeding this cow? Are we not breeding her?"

As Cool Breeze has always been a family farm, wholly dependent on family members working to sustain it, the early years for Niles and Esther were very busy times. Each had many tasks to finish by the end of the day. While the children were still young, Esther had a wider variety of jobs to perform during the course of a year:

I used to do all the harrowin'. I used to drive the silage trucks. As far as mowing and planting like

Plate 15: Linda Miner

that, he usually does that because that's not foot work. He usually has always done all the machinery work other than harrowing. When they would be cutting corn, as soon as they would finish a field we could put the rye on. I'd have it harrowed in and keep up with them. We used to harrow everything in the spring before we planted. Everything was harrowed, and then we harrowed all the rye in so there was a lot of harrowing to do. But now we don't do any spring harrowing at all. You just go over it lightly when you put the rye or wheat on in the fall, and the last couple of years he hasn't had them put it on until he's through chopping and that way he can get to put it on, so I haven't harrowed. I harrowed last year when he was in the hospital – those three fields we put sorghum in down there.

Much of each day is routine, although few days go by where there are no interruptions to the pattern. While every person working on each of these farms has their own way of doing things, one thing they share are early mornings and long days. Linda's schedule is no exception:

I got up about five o'clock this morning. I start milking the cows while my sister cleans out the feed bunks and gives them feed. Then she comes in and does the morning milking. I clean up and give them hay and so forth. I get a little break there usually. I work on paperwork or something, and then we do chores again where I water-up and take hay around and feed the heifers. Dad usually scrapes the cow yard and then you might get a little break, or you might be doing other chores. Today we had a heifer calve. I have a goat thinking of kidding. Afternoon chores start about 3:30 – getting things sanitized and ready to go, and cleaning out the feed bunks. I do the afternoon

Plate 17: Niles finishes mowing the first cutting of hay in the upper field.

milking. I finish with that about 7:00. Mom tends to the calves.

Esther's day begins a little earlier than her daughter's:

My alarm goes off at four o'clock in the morning, seven days a week, three hundred and sixty-five days a year, and that's three o'clock when they change that stupid time. Then I wait for Patricia to come up. As soon as she comes, she starts cleaning out the bunks. Linda starts setting up to milk, and I start sorting the cows and putting the cows in. As soon as I get them in, Linda has got things ready – she can start milking. By that time Patricia is through cleaning the bunks and I'm there to tend to electric fences and so forth while Patricia feeds the cows, and that's how we start the day. Then, right now that it's winter, I come in for about a half an hour, and then I go back and put the next batch of cows in. When it gets summer, and it's light out, I don't come back in. I go pull some weeds or do something for a half-hour and then go back and put the cows in. In the summertime it is light enough I can start my calves, but now it's too dark. As soon as it gets light, after I put that second batch in, I go tend to how many calves are in the hutches; I lug them down milk after I warm it if it's cold. I've got another batch of calves out there right now. I've got eight in it. I've got four in the hutches and eight in that batch. I have to give them their grain and their hay and clean out their building. Right now I have to make two trips down with milk because there are so many calves. Then I come back and I wash the bottles and come in the house and by that time it's time to get breakfast. We have breakfast, and then I can clean off the table if I'm lucky and get the dishes in the dishwasher. Every day is different. Some days I'm cleaning in the barn, some days I'm doing whatever. This morning I went back out, and Patricia got some wood shavings and we bedded the calf shed and the hutches, and that's the way it goes.

The afternoon milking starts at four o'clock. I usually go out, about quarter-of, half-past three, quarter-of four and start. If I have milk that needs heating, I get that on to the heat for the calves, and then I can go down and feed the first batch of calves their grain, but actual milking we try to start about four o'clock. I get through after I put the second batch in, if I have got the calves done ahead of time – then it's not too much longer before I can come in. But this is winter. In the summer I don't come in, because I mow lawns in the evening. I work in the garden and things like that. In the summer I don't come in until it gets dark. I run in and get him his sandwich for supper and then I stay out until dark. And then I come in and I go to bed, and hopefully I sleep! Not always, but hopefully. I plant a big garden. The garden is mine. Down beyond it's all vegetables. I've got peas in, I've got spinach, I've got lettuce. In fact, I planted my peas March 10th – the first ones. One year I planted them February 28th. It was a nice year, and they came up and they were good, and today Linda picked me up some onion sets and some potatoes, so I guess I'll be doing some more planting. I do a lot of canning and freezing and making jellies and pickles as well.

Over the course of the past two years, Niles has had several operations on his legs, which severely restrict the amount of work he can do. Because of this, he has been limited to doing the machine-work required on the farm, including the tasks he has always overseen – planting and harvesting, as well as other jobs such as removing the manure from the barnyard. This inability to do all that he has done for the past sixty years or more has proven to be a tremendous frustration for him, as he belongs to the generation that could do and did practically everything that had to be done on the farm.

There is not a machine on the farm Niles can't run or repair, and every other farmer in town has praised his unmatched abilities as a dairyman. He was as close to self-sufficient as any farmer could be. He could saw his own timbers to build his own barns or sheds with his own mill. He can diagnose and fix problems with any of the dozen or so tractors used around the farm and the various pieces of equipment attached to them. He built the roads that connect the fields he cleared as well as the walls that surround them. While the daily chores with the cows are generally left to the three women, he handles the field and crop work that will feed the herd throughout the year. He has kept abreast of the countless changes that have taken place over his many years on the farm, from crop genetics to the modern methods of planting them, updating and improving to keep the farm viable where most others have failed. Everywhere you look, in keeping with his self-sufficiency, are labor-saving tools that allowed him to do the work on his own. From welded tines that convert his backhoe into a forklift to the large steel workshop where fabrication and repair can take place out of the weather. In an age of specialists, Niles stands out as a farming Renaissance man, and few are spoken of more highly in the community.

Plate 18: Patricia Palmer

A major change in dairy farming has centered on the barns they use. All of the farmers in town began with stanchion barns. These types of barns, used commonly throughout New England, were generally built with a central isle running between two rows of stanchions. The milking cows were held in place in their stalls with wooden or iron pipe stanchions that closed around their necks. They had feed troughs in front of them, and milking took place a few cows at a time. Niles describes how this worked:

They had old Empire milkin' machine. I don't know, he may have

Plate 19: Niles spreading manure on one of his cornfields.

gotten McCormick by then. They just run off a vacuum – just a vacuum line with petcocks. The pipe ran right down the back of the stanchions, with a petcock between every other cow. We had milkers that milked two cows at once.

While the early milking machines were certainly a welcome change to milking by hand, there were other inefficiencies inherent in the stanchion barn that new designs would mitigate. While each farm in town still has and utilizes their older stanchion barns, they have all moved their milking herds into free-stall barns. The free-stall barn concept began after WWII. Rather than having every cow held in her own stall for feeding, watering, and milking, the new design allowed cows to move around the larger barns. They can wander at will, getting feed when they are hungry and water when thirsty, returning to any bunk they choose to occupy for rumination. History has shown that happier cows (or less stressed ones anyway) do produce more milk. These structures are generally much more open than the older wooden barns, whose only light came from doorways at either end or from electric lights hung above. While they have sidewall curtains that can be dropped for winter months, they remain open for the remainder of the year. This fresh air and sunlight can dramatically reduce health problems caused by keeping animals in confined quarters. Beyond the benefits to the health and well-being of the cows, there are significant time saving-features for farmers utilizing this type of barn.

Manure has long been a costly material to handle. While it forms the major component of all the fertilizing that occurs on these local farms, it also takes significant portions of the farmer's day to remove it from barns, store it, and eventually spread on their fields.

The advent of the free-stall barn revolutionized this tedious process. Free-stall barns are designed with the modern skid-steer loader or tractor in mind. Aisles are left wide enough for these machines to be able to travel through them on their concrete floors. Cows will generally enter the bunks where they rest head first, leaving the other working end of the animal and its by-products in the aisle-way. Three times a day, Niles will scrape the free-stall barn clean of the accumulated manure using his skid steer. The cow-yard is scraped once a day. As with the manure from the barn, it goes directly into the manure pit Niles built at the end of the yard. As one would expect, doing this work with a machine saves both time and energy, allowing the savings to be consumed by other tasks in the daily routine. When it comes time to spread the manure, it can simply be pumped or bucket-loaded into the spreader for distribution. As an added benefit, cleaner barns lead to fewer health problems and an increase in milk production.

While the cow's manure forms the backbone of the fertilizing plan for the fields surrounding the family farm, they must also buy some commercial fertilizer. Linda explains the manure-spreading process:

We scrape the manure into the manure pit where we can store it for about six months – until we can get on the fields and spread it. We didn't get much out in the winter this year because of the weather, so now we are spreading some of the solids. Last year we had to spread it almost every day because our pit was full. We try to give the cornfields a good fertilizing in the spring. Our pit is a semi-solid, so we do have a little pump that we pump water – the rain that's been in there – but after it gets down we have to load it with a bucket. Fertilizer we buy goes on the corn and the hay ground. The fields are measured as to what they need. Mostly it's nitrogen that goes on. The fertilizer company samples it and sends it to UCONN for analysis.

Plate 20: Always curious, a Holstein stops to pose for the camera before she takes a drink.

put quite a little on the corn this fall – put the cover crop on and spread manure on top of it. The corn fields get mostly manure here, but up on Pendleton Hill I have to put commercial fertilize on it. It's not convenient to haul manure up there. It's wet up there in the spring and I can't get on it. I have put some chicken manure up there years past, but haven't since I've been no-tillin'.

Everything now is no-till. Patricia harrowed a little bit up here last year – thought it might cut down on the micro-toxins, but I don't know it if does or not. I think that sometimes you have it, sometimes you don't. Where you harrow it up you get more erosion, and I've had good luck with the no-till.

Niles put up his free-stall barn in the mid-sixties. After having ordered the steel structure in the summer, he had to wait until the winter to have it erected. When the man who was supposed to put it up arrived with little help, Niles decided he might have to jump in if he wished to see the

The manure pits allow each farm a little more flexibility in spreading the material. In the past, most manure would be spread almost immediately to keep it from filling the barnyard, reducing its effectiveness as rain and snow would often wash away the nutrients it provides. The manure pit gives them a chance to spread it at the best time of the year, although as Niles points out, this is not always the case:

benefits of the new barn before the end of the upcoming winter:

Generally we don't spread too much manure in the winter. The last couple of years we haven't been able to get on the fields in the fall because it has been so wet. I used to it on the grass in the fall, but I don't want to cut the fields all up. I

So the contractor comes down with a couple of kids and asks me about taking the loader and setting the rafters. He had bolts in the foundation that the posts set on. I said, "I'm not through milkin'." He said, "Oh, I don't know if we'll get these posts set up today or not." I was younger then and I could handle those posts alone. I would set them down on the bolts – all they had to do was put the nuts on. They weren't makin' very fast progress, so I said, "You mind if I try and get a little help?" He said, "Oh no! I'll pay them." I don't know if he ever did or not, but anyway, I called my cousin George up on the hill. He put the purlins in I guess, or the girts right along the edge – something to step on. It was all drilled – all we had to do was put a bolt in and put the rafters up. I think we had the purlins sittin' up there that night. George would go up one post, and I'd go up the other and just stick the bolts in and catch the nut and move on to the next. I think those kids were three days tightening bolts. Then the contractor got sick – didn't come for a week – took the blueprints with him. Anyway, we had it pretty well closed in before we ever saw him again. Then he had to get the doors and the partitions and so forth. That winter I left the cows right outside. They'd go up in the woods, in the snow, and they milked better than they ever did in the barn. But we finally got in. I know it was snowin' like a son of a gun, and I went down there and got a little gravel and some shavings to put in the stalls. They weren't filled like they should be, but at least the cows had a place to go in. That was just before my father died. The cows get a little freedom – they can come or go or whatever.

Along with the advent of the free-stall barn came the modern milking parlor. Even after the first automatic milking machines became popular, farmers still had to go to the cows in their stanchions and sit on a stool to milk a few at a time. If not piped to a collecting tank, they might also have to carry the milk to a central storage container. The milking parlor changed all of that. The parlor is built with a recessed pit in which the farmer now stands. The cows enter the parlor on both sides of this lower alley, and the depth of the pit is such that those milking the cows now have the cow's udders right at shoulder height, eliminating the need to constantly bend down and stand up while working with the animals. This gives them a better view of the cow's underpinnings, so not only can they work faster and more easily, but also see if there are any health problems developing with any particular animal. Niles put in his parlor in 1963:

The first day I used the parlor was the day my mother died. Originally it was a DeLaval parlor. The dealer used to come around from East Providence and I bought it from him. It was a small parlor – a double-three. I'd watched Woodmansee milk a couple of times. He had, I think, a double-six with two people in the pit. Of course he had a lot of cows. I said to myself, "Three is about all one person wants to tend to." When my cousin Franklin sold out, I bought the parlor that he had. He had indexing rails that crowd the cows and so forth. I decided to go with a parlor because we had to do something – the state said we couldn't milk on a wood-floor barn anymore. I think this was the first one in town. Of course most of the other farms had cement floored barns.

Plate 21: Patricia begins the morning round of milking in the parlor.

Cows today are trained to enter the milking parlor with some slight prodding. Esther's first job of the day is to separate the cows into groups that will be milked together:

We usually put all the first calf heifers – we call them heifers – in the first group, and then, if there's a cow that has just freshened, we might put her in the first group, or if you have a cow that's giving a lot of milk and is a little timid you might put her in because they're smaller cows in the first group, but it's usually heifers in the first group and cows in the second group. The heifers – most of them are a little timid when they first start – so that's my job to go out there and sort them and put them in. I have six that I'm putting in the very first batch – put them right up in the parlor. They know when I'm coming, and some of them will head for the door and some of them will head off in the other direction. But they usually sort out pretty good.

Parlors are rated by the number of cows that can be milked at one time. Cool Breeze has what is called a "double three" parlor – that is to say three cows can be accommodated on each side. A gate is opened and a single file of cows moves into the parlor on each side. When the appropriate number of cows are in the aisles the gate is closed behind them. They are "crowded" into position for milking. The cows are not held in place by stanchions, but rather know to stand at a milking station with their udders to the pit and their heads to the outside walls. Their hips rest against rails at each station as they are prepared for milking. Linda describes this next step:

Plate 23: Niles scrapes manure from the yard into the manure pit.

You prep the cow, you dip it – our dip contains some peroxide. It's to kill the bacteria on the teat. You wipe the teats and get them clean and dry and put the machine on. The machine does the work, and when she's done milking you take it off and dip her again and she can go out.

Plate 22: 1955 Ford 800 tractor.

Electronic boxes monitor the pounds of milk taken from each animal and allow the farmer to see at a glance how each animal is doing. If desired, these boxes can be hooked up to a computer to track each animal's history, but currently none of the farms in town go to that length. As they have all stated, working with these animals as much as they do every day, they know them pretty well. The cows remain standing at their sta-tions until the milking for all the animals on a side is done. After re-dipping the teats to help prevent infections, the gate at the head of the parlor is opened and the cows walk back out into the yard, where they can resume doing whatever they would like to do – eating, sleeping, or standing around the water cooler.

Cool Breeze Farm currently milks between fifty and sixty cows. This number may vary on a weekly basis, depending on which cows are calving or about to calve, and where they are in the milk-production cycle. According to Patricia:

The herd has fluctuated a little bit. I can remember at one point, after I moved back here, we were milking around seventy, but that was too many – too crowded and took too long to milk – so we're back down milking in the fifties again, and I think that's just about right for the number of stalls we have and the amount of cropland we have, for the feed that we get. It seems to work out pretty good to keep it right around fifty – between fifty and sixty. From my perspective, from being in the milking parlor, that's enough. If I'm in there more than two hours, I've had it – I want out of there.

The number of cows they milk is by no means a complete picture of the overall herd size on any dairy farm. It takes approximately two years to bring a cow from birth to the milking parlor and cows are leaving at the far end of the production cycle throughout the year as well. Linda states:

We have fifty-six milking with the heifer that just calved today, or there will be fifty-six. We have three waiting in the wings right now. We try not to have too many calve in the winter. Young calves – there's three in the hutch, there's six in the small group, like eight in the middle, a dozen out back, a dozen across the road, I think there's five down to my sister's house, which they will calve shortly. We usually have about twenty calves, new cows, every year.

Plate 24: Linda heads to the field with the hay wagon to pick up the newly dried and baled hay.

The major change in farming throughout New England, in terms of outlook, has been the switch from subsistence farms, which also created income by selling any extra milk produced, to the operation of a dairy farm in which the majority of the time is spent creating a product for consumption off the farm. While milk cows were never entirely a casual pursuit in the early years, the change to making dairy the means of a living focused the effort that went into milk production. There is nothing casual about the modern dairy farm, and a surprising amount of science goes into each gallon of milk shipped.

When asked about significant changes she has seen over the years she has worked on the farm, Linda had this to say:

I think that's probably the biggest change in my lifetime – how much nutrition information they've figured out on a cow. They balance every micro-nutrient and everything they can think of for the health of the cow. We feed grain, and we have a molasses mix – sugar mix – that we put on the feed, and that's about it. But the grain is pretty expensive. It's usually about five tons per week that we get for the milking cows, and then we get about two tons of grain for the heifers, but that goes for three, four weeks. Right now that's just a regular sixteen percent grain, whereas the milking cows, the grain is specially formulated for them based on the other feed that we feed.

The grain company takes samples and analyzes them. They provide the nutritionist that tells you, "You need to feed more grain!" It's pretty balanced, but my sister will talk to them and say, "Well, you know they're loose, there's a little too much energy, or they won't eat it. Let's change it." You work with the nutritionist to get the right balance. Sometimes, something that looks good on paper, the cows will say, "Nope – no good!" I remember once – it was a different grain company that's no longer in business – but somehow they got too much bi-carb in the grain, and you'd mix it with the acid in the silage and it would stink, and the cows would not touch it. You could offer them the ingredients separately, you could offer them the grain separately and they would eat it. But you mix it, and they would not touch it, and it stunk, and

Plate 25: View down the hill into the barnyard.

tance for the farm not only to run as lean as possible, but to produce as much milk per cow as possible without overtaxing the animals. Patricia explains this nutritional balancing act:

So it's a fine line between giving them as much grain as you can get into them to get as much milk out of them without causing the health problems, because then you're going to end up losing in the long run. It's always a balancing act. You want to get them right to the edge to get as much milk out of them, but don't push them over the edge, otherwise you've lost! Then they start having foot problems, mastitis problems, and all sorts of metabolic problems. The way the economy is, if you don't maximize as much milk per cow as you can in a small operation, you can't be competitive. The way the economy is now you can't make any money on it, but to make the most amount of money, at least from our

Plate 26: The old milk house.

point of view, you've got to maximize your milk production, 'cause that's the only thing that's bringing in that paycheck. On the other hand, you can't push them to the point of death loss, lameness, or vet bills.

As with all of the farms in town, the nutrition animals receive is based on the advice of a nutritionist in conjunction with the thoughts and observations of the farmer working with the cows. The investment in each animal is too great to leave to trial and error. Patricia explains:

I don't personally sit down and calculate the ration. I work with the nutritionist and give him my observations if we make a change – I tell him this happened or that happened. It's so detailed now. You have to do it with a computer. You can't just sit down with pencil and paper and figure out the ration anymore. It's incredibly scientific – it's just incredible the number of things they look at. So I work with the nutritionist from Cargill Company. That's who we buy grain from, and he's really good.

So Craig is out there – he's looking at the cows, looking at the feed, looking at the manure. He tries to look at the whole picture. My mother will give him some input, Niles will give him

it turned out some computer problem had put in too much bi-carb, and we ended up shoveling all that out and getting something different. The cows today aren't even the same animals as they were fifty years ago. Fifty years ago you turned them out in some brush and said, "Go to it!"

With today's higher production levels of milk and thinner profit margins (or as was the case for the year in which these photos were taken – negative profit margins) it is of the utmost impor-

some input, but generally I'm the one that goes out with him to look at the cows, and I go over the dairy-herd records with him – you know – what happened with production after we made a change, or the butterfat or protein – did it go up or did it go down? I relay what the vet says when we do our vet checks. I always ask the vet, "How's the body condition?" The vet sees the cows maybe once a month or once every two months and it's very easy if you don't see them every day to pick up on a change, whereas if you're there every day, cows can lose body condition so gradually you don't really pick up on it. So I'll quiz the vet and get his input on what he observes when he's here and then talk to Craig about that, so he can incorporate that into any changes in the ration.

When we first open a new pile of haylage or a new pile of corn silage, we'll try to test it as soon as we get it open to get a preliminary picture of what the feed is, because it could be very different from what we were previously feeding them. The weather from one year to the next can have a huge effect on the feed. After you've been feeding it for a while we generally like to re-sample again because the longer it's in that pile fermenting, the feed analysis will change, so we don't just use that one initial report. We tend to, once you get feeding it for a while, re-sample again. And then if you notice that the cows just aren't doing what they should be on it, then we'll say, "Well, it looks like maybe we need to sample it again – maybe something has changed that we're not aware of." We sample the feeds several times a year anyway, even though it's the same pile of feed.

Milk production is measured and sold on the farm by the "hundred weight," or one-hundred pound units. A cow's production by day is also by pounds of milk rather than gallons. Niles compares the difference in production from when he got married to today:

The genetics with the cows have changed quite a bit. Well the feeds have too. I don't know if it's better or worse. Price-wise it's worse. We got 50-60 pounds of milk per cow when we married, and today over a hundred pounds. Well, we only used to feed hay and silage, a little grain, probably a sixteen percent ratio. They test the silage, grasses, and so forth now, and try to balance it somehow with the grain.

The number of pounds of milk any particular cow will produce per day as well as the herd average are also highly variable. Patricia notes:

That's sometimes hard to judge, because the pounds per cow will change if you have a lot of cows freshening or if you have a lot of tail-end cows. A cow's production cycle goes up and then tails off, so the days in milk that the herd averages is going to vary. There's a lot of variety just because of the age. A first-calf heifer isn't going to give anywhere near as much milk as a cow that's in her third or fourth lactation, so there's going to be a lot of variation there. Some of the heifers are dogs. You know, you take your best shot. We don't raise all the heifers – we kind of have to pick and choose and you just take

your best guess, but until they actually calve and start milking you don't know what they're going to do. Some of them are just dogs, but then some of them will out-milk any others. There's some individual variation, but as a rule they're close. The heifers will be within a certain range, give or take ten pounds.

Linda explains the typical life cycle of a cow on Cool Breeze Farm:

Plate 27: Patricia cutting hay as the corn begins to grow in behind her.

We usually separate the calves at birth from their mother and we feed them on a bottle. At about eight weeks they're weaned, and then we have different pastures, based on the age of the calf, which they go into. They're bred to calve at about two years of age. You try to shorten that a little bit if you can. It depends on the size of the animal, because until they start milking, they're costing you money – it costs you money to feed them every day for about two years. Once they calve they start milking, and then you have to breed them so that they calve every year to continue milking. They milk for ten months, and then they get a couple of months for a dry period, where they aren't milking. It used to be, a long time ago, where the herd would calve in the spring, but nowadays it's a continual thing because you need milk continuously.

Linda further explains how the cow dries-off before the birth of her next calf:

She pretty much does it herself. She slows down producing milk. We do treat them with antibiotics in the udder for protection, so they don't get bacteria, mastitis up in there, and she's put into another pasture, not fed quite as heavily for about two months. Then she calves and she's back in with the herd, usually as soon as she starts producing milk again. Usually that first milk is fed to the calf, and we test the milk to make sure there are no antibiotics left in it, and then it will go in the tank.

Plate 28: Niles chopping corn. The dumping wagon behind can lift to fill a dump truck with the corn.

The animals' feed ration is also determined by age. The animals are roughly divided into five groups – calves, heifers, far-off dry cows, close-up dry cows, and milking cows. Heifers are cows that have never calved, and therefore produce no milk. They generally enter the milking phase at around two years in age. The milking cow's ration is not the same as the heifer's coming up into production, and of course, the newborn calves' needs are vastly different from more mature animals. Esther is in charge of feeding the youngest of the animals from birth and into the heifer stage:

When they're first born, they get the cow colostrum. And then they go onto the regular cow milk, and they get two bottles twice a day. The bottles hold two quarts apiece or a little bit better, so they get eight quarts a day for anywhere

from seven to eight weeks. In the winter I try to keep them on eight weeks. In the summer, when you've got real nice weather, I figure seven weeks. Then it takes another ten days to two weeks to get them weaned down. I don't just cut it right off. After the two bottles, then maybe I'll give them two bottles once a day, and then one bottle once a day, and I keep gradually weaning them off for a few days. They don't eat an awful lot of grain when they're on that much milk, but as soon as you start taking the milk away, then they'll eat more grain. I get them used to the hay then. Then, when they get eating hay and grain, out they go into the other pen, so I have them for about ten, – nine, ten weeks in the hutches. We go sort of by when we have room. If this batch out here by the road is full, I can't move mine over. I think there is one more open space here so I can move one more over, and we have a cow that's thinking very strongly of calving today. So if it's a heifer, I'll have one more out there. We've had quite a run on heifers. We didn't for a long time – we had nothing but bulls. Lately we've been having quite a run on heifers and everything gets filled up in a hurry.

Cows are ruminants, with a digestive system quite different from ours. Patricia explains how cows digest their food:

A rumen is just a large fermentation vat. There's bacteria in there. That's why a cow can eat hay. You can't eat hay – you can't live on it because you don't have that fermentation vat in your digestive system. They say a cow has four stomachs. Well, it doesn't really have four stomachs, it has these four compartments. There's the rumen, where the bulk of the digestion occurs, and yet when they're a calf they really don't even have a rumen, they have more of a true stomach. There's a rumen, the reticulum, the omasum, and the abomasum.

Plate 29: Corn getting its start in the early spring.

When they're a calf, the omasum is what pretty much digests the milk, and that will play a smaller and smaller part in their digestion as the rumen gets bigger and bigger. And off the rumen is what is called the reticulum. If you've ever had tripe, that's the reticulum. When cows eat, they don't pick at their feed – they just engulf their feed – and then they go lay down and chew their cud, and that's when they're chewing it to start the digestion. When cows were developed, one of their survival mechanisms was that they would go out as a herd and eat as much as they could eat in as short a time, and then go where it was safe and lay down and chew their cud.

After they chew it, there's a mat of feed in the rumen – a lot of fluid in there too – and the bacteria

are actually breaking down the feed, and then it gets small enough to leave the rumen and then go on down to the abomasum, which is the true stomach – the same as your stomach and my stomach and a pig's stomach.

It leaves the abomasum and goes on through the intestines and gets further digested. When cows get what they call a twisted stomach, you can detect that with your stethoscope by

Plate 30: Corn at mid-summer.

listening to the rumen and the pinging. A healthy rumen will sound almost like a toilet flushing. You won't hear much and then you'll hear the sound, and if you're holding your hand on it, you can almost feel it, and you can definitely hear it with the stethoscope. Sometimes you'll get cows that are off feed and you can listen and the rumen will be absolutely quiet. There's just no rumen movement, and if there is no rumen movement they're not digesting their feed. You can detect a very sick cow that way.

Calves are not able to start with feeds given to older cattle. Patricia further explains:

The little calves that are in hutches get the milk and then a calf starter grain, which is a grain designed to really stimulate the rumen development, because they really don't even have a developed rumen when they're born, and it takes a while for that rumen to develop. And hopefully by the time they're weaned, they're off the milk and then they're on grain and hay completely, you've got that rumen so it can handle those feeds. So those little calves, they need a higher concentration of grain in their ration. As they progressively get older and they can eat a larger percent of forage in their diet, the grain will get cut back. As they progress from one group to another, you're gradually changing their diet a little bit, and by the time they get down here to my house – these heifers are the next ones that will be calving – they've been checked pregnant – and when they get close to calving then they go from here back to the milking herd. So you've got to take into account, "OK, so these heifers have a fairly large fetus inside." Your feeding not just the heifer, but the calf too, and you're getting them geared up that they're going to be having to start producing milk, so you have to change their diet a little bit. And then with the milking cows, it's just a one group TMR, (total mixed ration) because we don't separate out fresh-cows and tail-end cows or heifers from the older cows. We don't have the facilities. Ideally it would be really nice to feed the new heifers their own little diet in their own little group. The fresh cows, their diet; the tail-enders, their diet. You can be a lot more efficient that way, but with fifty cows you can't do it. It's one diet – you're all together, you're all going to have to eat the same diet. High energy, but probably not as high protein as what the little calves are getting, but certainly higher than what some of the middle-sized heifers are getting.

Plate 31: Patricia sets a round hay bale on the wrapping machine.

Plate 32: Linda wraps a round bale for storage.

Energy is basically the limiting factor on how these cows will milk. You can easily get enough protein in their diet, but it's the energy that will limit them. So the better quality silage you have, the better quality haylage you have, you get more milk because you're not taking up rumen space with stuff that they're not getting the energy out of.

You look at energy, you look at protein, and then you look at the minerals and the vitamins, and energy is basically your limiting factor on how much milk you're going to get out of those cows. You want to put the best quality forage into them, because you can only put in so much grain before you're going to be causing health problems.

They're ruminants – they're designed to thrive on forages, not grain. Like a pig – a pig isn't a ruminant – you can really poke the grain to a pig and they'll do fine, and the chickens, but with cattle, you go poking the grain to them and at some point you're going to have problems.

While attention to feed has certainly benefitted production levels as farms have moved forward in time, it is not the only thing that has altered in the life of a dairy cow. Genetics, in both the animals themselves and the crops they eat have also evolved. The change in the appearance and qualities of the modern milking cow has been rapid. According to Patricia:

You try to make a more efficient cow. If you look at some of the old-time pictures of cows and you look at the modern-day cow, there's a huge difference. I have some old USDA year books – they used to put out an agricultural yearbook every year – and you look at some of the old ones and the pictures of the modern-day milking cow, and it's amazing, the difference in their appearance.

Dairy farmers now spend a lot of time and research effort on designing the cows they want in their herd. While they still use bulls to impregnate cows under certain conditions, most of the breeding is now done with AI, artificial insemination. This gives the farmers a finer tool for improving the traits of the herd as a whole. With AI, each farmer has access to the gene pool of bulls from around the country, not just to the one standing in their field. As it takes nearly three years to bring a cow to the milking stage (nine months in the womb and two additional years before they themselves calve) and all of the expense this entails, knowing what you will get is an obvious benefit. As Patricia notes, animal genetics have come a long way:

Everything now is so sophisticated, with all of the DNA mapping in cows. They can just take a blood sample of a cow or a bull or a calf and now predict what that animal's production will be. It used to be all these AI organizations had to raise up a bull calf, based on the parentage. They would go out and seek out bull calves they thought would produce good offspring, but they didn't know until they were tested through these progeny testing programs, and now they've got it where they can just take a blood sample on the calf and they can see what that calf inherited for genes from its parents and make a really good estimate on whether that animal is going to make a good bull or not. They still go through the Young Sire program – having it sampled. They'll put the semen out on the farms and get the infor-

Plate 33: Fog shrouds a field. The cast-iron tubs next to the feed bin serve as water troughs.

mation back once the calves are born and they start milking, but it's amazing how they're shortening the interval. Because it would take a long time, from the time a bull calf was born, the bull studs would buy the calf, take it, they raise it up – well it's got to be a couple of years old before they start collecting semen from it – put it out on a farm, then it's at least another nine months before a calf is born. It's going to be another two years after that before that animal starts milking and then another year to get a complete lactation, so there's quite a delay. Now they're shortening that up by the DNA testing on these animals and rejecting some without even putting them through the testing program because it didn't inherit the right genes from the parents. So there's not even any chance that they could pass it on to the offspring since they didn't inherit the genes from their parents, so they just eliminate them from the testing program from day one.

Patricia is in charge of the breeding of the cows on Cool Breeze Farm and actually breeds them herself after having learned how at an on-the-farm AI school. Semen is purchased from an AI company and stored in liquid nitrogen until needed. There are any number of AI stud services available to the local farmers and each farm has their preference, based on what they imagine the perfect cow to be. Patricia explains:

You're not locked into just breeding your cows to one or two individuals. You can completely tailor your breeding program – what is the best bull for this particular cow – and you've got hundreds of thousands to choose from. It used to be I'd sit down with the AI book and try to match them up, but now I just work with an outfit – they come in and do the mating program. He looks at all the cows and analyzes them for their physical features and I give him access to the DHI records which gives their production records. So he gets a snapshot of each individual cow and then can say, "OK, your best bet is breed this cow to this bull – your first choice, your second choice, third choice."

The AI companies' books on bull semen list numerous traits possessed by each bull. As hoofs and legs are vitally important to the cow, and in fact are often the lead indicator when something goes wrong health-wise, many farmers stress that they try to improve on these qualities every time they breed their herd. Good udders, milk production, daughter pregnancy rates and longevity are other key components considered when picking the right bull for the job. Patricia, like all farmers, knows what she is after:

Well, these AI studs have their own philosophies, they all differ a little on the ideal cow. The old Eastern cows were short-legged, big-uddered – they would out-milk anybody else's cows, but they wouldn't last. Their feet and udders would give out on them. I can remember milking these cows. You had to be a contortionist to get the machines under them, because the udders would be so low to the ground. I don't think there's a machine now on a cow that hits the grate. We've gotten the cows up in the air, the udders up in the air. This mating programmer I work with – he comes twice a year – and I ask him, "After you get everybody put in, give me a printout." They look at, let's say, two dozen different physical characteristics, and I'll say, "Give me a printout showing me where we are – where the herd average is." So I've got it from day one, 'till now, and what a difference. I just wish I had pictures of those cows, because in my mind I can kind of picture them, and I know what it was like to milk those cows and I know what it is like now, but I wish I had it in black and white. I have a graph which shows a dramatic change.

The initial decision may simply be what breed or crossbreed will make up the farm's milking herd. While Holsteins are the predominant dairy cow found in herds throughout the United States, there are reasons why farmers may choose another. Patricia:

There are five major breeds. The Holstein, Brown Swiss, Ayrshire, Guernsey, and Jersey. Holsteins will typically, as a breed, produce the largest volume. Because it's the largest volume, it's typically the lowest in components. The components are butterfat and protein. The Jerseys are at the other end of the extreme. They tend to give the least amount of milk, but it's the highest in components. The others fall in between. Jerseys I think are making a kind of a comeback now, because for a while there people wanted low-fat. Everything was low-fat, no-fat, skim, and to some extent it is now, but cheese production, cheese consumption, has taken off. You're going to get more cheese per pound of milk from

Plate 34: Linda spreads straw as bedding in the calves' open-air shed.

Jersey milk than you will Holstein milk. So if you're trying to capitalize on the market for cheese production, you can get more per pound for Jersey milk than you can Holstein milk, so the Jersey has kind of made a comeback. That occurred a few years back. The pricing – it used to be you didn't get paid anything for butterfat and protein, and then, all of a sudden, protein and butterfat is worth more than just the volume of milk, so Jerseys started becoming popular again. We have one. There are some farms that do run a group of Jerseys and a group of Holsteins.

Patricia has kept the Cool Breeze herd closer to pure Holstein, although even among her family members there is some difference in opinion on what animals they should have. Her philosophy on this matter is clear:

If you want to go for components, then have the Jerseys. If you want to go for the volume, have the Holsteins. When you put the two together you get an animal that gives you more components, but less milk, and then what do you breed that crossbred to? That's the big stumbling block to these herds. A few years back, that was the big thing – going to these crossbreds. That's great for the first cross, but then what do you do? What do you do with that animal? She's a little too small to breed back to Holsteins, so OK, we're going to breed her to a Jersey, so eventually you're ending up with Jerseys. They just don't fit in a free stall that is designed for Holsteins. The bunks are designed for Holsteins, the parlor is designed for Holsteins. You stick in these little animals and they can't compete. You've got a thousand-pound animal and she's supposed to compete with an eighteen hundred-pound Holstein?

As previously mentioned, the herd's total milk production depends on where the herd is at the moment in the calving cycle. As each cow produces more or less milk at either end of the cycle, having them all on the same cycle can lead to real fluctuations in the pounds of milk produced at any given point in the year. While farms may try to flatten this roller coaster by breeding throughout the year, this hasn't always worked on Cool Breeze Farm. Patricia:

Our herd is very cyclical in calving. We calve a lot late summer, early fall, and at this time of year [March] we have hardly anybody calving, so the days in milk of the herd, the lower the days in milk, the higher the milk production per cow per day. So right now, we're on the way out in days of milk – they're getting further and further out in days of milk, so production per day per cow is dropping. We're right around eighty pounds per cow per day. In the late summer, early fall, when we have a lot of cows freshening, that should go up, and as we get further out now in days in milk it should be going down. Some herds, their days in milk probably stay very consistent. For some reason, this herd has always been a fall-calving herd, so to speak, ever since I've worked with it. Why? I don't know. I don't consciously aim for that. Years ago I know they did because they got a bonus. In the spring is when a lot of the old-time cows would calve – animals calve in the spring – they're not genetically designed to calve in the fall, so there was a flush of milk in the spring, there was a shortage in the fall, so the milk companies would pay a bonus in the fall, less in the spring, so it was to your advantage to have more cows calving in the fall. Whether they got the genetics set up that way or just got lucky I don't know. For some reason, we don't calve an even number throughout the year.*

While she hasn't been able to control all of the calving the way she might prefer, she does have several goals in mind when breeding first-time heifers:

The only time I try to manipulate when they calve is the very first time the heifers calve. It's hard on them, January, February. Heifers will get edema, and if they're standing out with wind chills, they get frostbite really easily. If it's borderline when to breed them, if I can

Plate 35: Without crops to tend, winter workloads are lighter, but slower in the cold.

Plate 36: Patricia pours an acid preservative into the chopper that will lower the ph levels of the haylage they are about to harvest.

get them to calve a little bit earlier than January, February, sometimes I will breed them a little early, or I might think, "I can hold off a month or two, calve them in March." It seems like January and February are the worst months. But sometimes you get a heifer – they don't always breed the first or second time – well once they get a little age you don't want to be holding off. You want that animal bred – get her in the milking herd. You can fudge a little bit, but not a lot. I have tried to do that a little bit on heifers, but on older cows I don't. To be most profitable you want to get them bred back as quick as you can. You don't want to be trying to avoid a certain month of calving.

The breeding of dairy cows has even progressed to the level of sexed semen, although several farmers have wondered about the unintended consequences this may bring along with it. Bull calves are of little value today, and the cost of shipping them often cancels out the little money farmers receive for them at auction. Beyond that cost, however, is the additional cost of feeding a cow for

nine months to produce a calf of no value, and thus the potential benefit of using sexed semen. While Cool Breeze did produce its first calf from sexed semen this winter, Patricia still sees long-term problems with the process:

It's more expensive, and the conception rate is lower. If you're trying to expand your herd and you want more heifer calves, it's a real good deal. If you have a good market for excess heifer calves it's a real good deal. I think it's like anything. Everyone is using sexed semen now – heifer calves aren't going to be worth anything a year from now. There is going to be a flooded market for heifer calves. If you're using sexed semen today, you've got to judge what's the market nine months from now, and that's kind of tough unless you know what price calves are going to be bringing, and the price of milk. You're just guessing. Right now there's no market for bull calves, and we do have a market for excess heifer calves, so it's an advantage. But the other side of the coin: it's more expensive, you get less conceptions – you've got to breed them a little bit more to get them bred. The big advantage to using sexed semen on heifers is you're not fighting these great big bull calves that are difficult to get out of the heifers, and that's worth a lot right there. The heifers that have a heifer calf are going to have a much easier time calving. That's a huge advantage and a good reason to use sexed semen.

Every cow has a limited working life. While some cows have made it well into their twenties, the modern working cow is done at a much younger age. Beyond production rates, however, there are many other reasons for selling off a cow. As Esther notes:

With the older cows, there are always some going, even if they have a bad disposition. They eventually, you know, get bad feet, or they get mastitis and foul up their

Plate 37: A contented cow in her field.

udders. There is always something that they keep moving on. And if Patricia doesn't want them after they've calved for a little while, she always makes a little note, "Do not breed" on her charts, so there are several that are "do not breed" on her chart in the milking parlor. And then there's one once and a while that's just plain old miserable. I've given her a list of a couple – get rid of those. Don't you dare breed them again!

We can only milk about fifty-five – a little bit more – sixty cows, or we run our tank over. So we're limited to how many we can keep, and that's good, because the building holds sixty-eight. I think there are sixty-eight beds, but we can't milk that many because the tank won't hold it. It's enough. It keeps them in the parlor long enough. You get tired of that after a while. You have the cow two years before you milk them, and you usually get five, six, seven years. They used to last a lot longer, but now they're pushed so hard that something usually goes before too many years. They never used to get a hundred pounds out of a cow in a day. A hundred and twenty, a hundred and thirty, some of ours. That's a lot of milk, and so something will go on them before they get to be old aged.

Plate 38: Niles chopping haylage in the field above the farmhouse.

that are pushing eighty, and we don't have any big, rugged guys to manhandle animals. You get an older cow that goes down in a stall and we're really stuck. We've got to have healthy cows, because we just don't have the manpower to deal with down cows or sick cows. If they start to show any signs of being a potential problem, I don't breed them – just milk them out. And I'm kicking myself – that old cow that was in the pen – she's getting better now, but I'm thinking why did I breed her back? She's a prime example. Do you want to fuss with that animal, or do you want to ship her out and replace her with a young animal that you don't have to fuss with? That's just the way we're set up. We don't have sick pens. We don't have pens for cows with special needs. It's a lot of extra work if a cow is sick or needs special attention.

Once the milk is produced, each farm must store it temporarily in refrigerated tanks until the trucking company picks it up each day. These trucking companies are separate businesses from the milk processors who buy the milk, process, and package it. Linda describes what happens to their milk when it leaves the farm:

We're paid by the hundred pounds. DFA picks up the milk, and they pretty much take it to whatever plant needs it, whether it's Agri-Mark or Garelick's, or Guida's, or some cheese plants. We used to say our milk is Garelick's, but now you don't really know – it can go anywhere. They go to all the other farms. The milk is tested at the plant. Say we treated a cow with antibiotics, we have a test that we can do right here before any of her milk goes in the tank so we don't get the antibiotics in, but that's pretty much what they test for when they take the samples up to the factory – make sure there's no antibiotics. They also test for bacteria. Tomaquaq does the hauling, and it does seem strange in this area, because they

Partially because of the limited workforce available to them at the farm, Patricia echoes her mother's thoughts on the subject:

We have a few old ones, but again that's kind of a personal philosophy. Do you hang on to the older cows longer because they've basically paid for themselves? But they're going to have more health problems – you're going to have to fuss with them. Or do you replace that older cow with a younger one who potentially has better genetics? And we're not equipped. We've got two people

will have three different trucks going to three different farms, but I understand they do that because these are small farms – they can balance the loads. When there were three different farms over in Clark's Falls, three different trucks would go there, and now we have two trucks going here and they follow one another. But that's why – they use it to balance their loads. It's the same company. There's no choice now – everyone's the same. They own the milk as soon as it leaves the yard, but we have to pay for the trucking. It should be changed, but I don't know if they will. I think they know they have a good deal there.

As the dairy industry dwindles in New England, so do each farmer's options. As Esther points out, they have few places they can now turn:

The sad part around here is there's only this one company that picks up milk. There's only one or two grain companies. You're stuck. You either go with them or what are you going to do? You have no choice. We used to have three different milk companies go right past here every day. You got ticked with one, you'd say, "Don't pick it up, we're going with somebody else." Now you don't have any choice. You take what they give you and that's all there is to it.

The smaller farms in town face other economic difficulties merely because of their size. Patricia notes:

There is such an economic advantage if you're large. Anytime that you buy any of your products, there is just such a tremendous price advantage the more you buy. It's really hard for us to compete. We're probably one of the smallest dairies in the state. You pay top dollar for everything, because obviously the larger you are, the larger the quantity that you buy, whether it's grain or any of your dairy supplies. The larger the quantity the less you pay per unit. So we get stuck paying top dollar for everything. You have to try to offset that by maybe doing a little better management job, getting that extra pound of milk out of the cows. But you can only get just so much. I think as far as efficiencies go, that's why everyone gets bigger, because there's the advantage of reducing your costs per unit. Bigger farms can get stuff a whole lot cheaper than we do because they'll buy a pallet load of this or a pallet load of that.

Linda points out another obvious disadvantage to being a farm wholly dependent on family labor:

We try to milk every twelve hours, without too much difference. Make sure that their stalls are comfortable for them to lay in, that they're fed right, that they always have feed in front of them. You have to keep all of the equipment running,

and then you have to find time to actually use it. You spend all day feeding the animals, and the other half of the day cleaning up after them. That's the hardest part about dairy farming. You can't just throw them some feed and call it a day, or put out a bale and go on vacation. You have to tend to them at least twice a day, every day.

When asked about milk pricing, every farmer has rolled their eyes and stated there is nobody who seems to fully understand milk pricing. Esther feels the same way:

I don't think any of us understand the pricing. They give you what they want and give you a sad tale and tell you that's it. They kept saying, "Oh! Milk's going to be up – milk's going to be up!" Now they're saying it's going to be down for a couple of months. We used to be in different sections, you know, the Northeast and the Midwest, and the Pacific coast, and now it seems like they're sort of lumping everything together. It's so much more expensive for the farmers here in the Northeast to make milk than it is for somebody in California or even

Plate 39: Heavy snow sweeps in, covering the fields and the old milk house.

in the Midwest where they grow all their own grains. We have to have everything shipped in, so our prices are sky-high, and they've got it right there. We should get twice as much for our milk as somebody that has the feed right there.

With modern transportation, the usual role supply and demand plays in the marketplace has disappeared. Huge farms from the west, milking tens of thousands of animals, can ship their milk anywhere in the country. New England farmers, whose costs to produce a gallon of milk are much higher, can lose out when flooded with the cheaper milk produced in areas where feed, labor, fuel, and many other costs of production are lower. This leads to constant fluctuations in the pricing of milk, making it even harder for New England farmers to keep their heads above water. This can lead to actions on the farmer's part that may exacerbate the problems they face. As Patricia points out:

I don't think anyone milks more cows because they want to – you milk more cows because you have to. I don't think anybody says, "Gee, we're having so much fun milking fifty, lets milk seventy-five! We'll have half again as much more fun" You get in that bind where you have to. You kind of get locked into it with the way the economy for farming has been. We are our own worst enemy. The more efficient you get, the less you're going to get for your milk. When the price of milk is good, a lot of guys put on more cows because "OK, now we can make more money!" Well, then when the price drops, they have to put on more cows because they still have to cover those costs, and you just get caught in a vicious cycle.

And all this new technology. When BST came out, it was the greatest thing. All of a sudden you could get so much milk out of your cows. Well, what did it do to the price of milk? The price of milk went down. Well, then you would have to put more cows on BST because you're not getting as much for your milk. (Note: Farms in town do not use BST.) *It's going to be the same thing with the sexed semen now. We're starting to see the effects of it, where there are going to be so many heifers out there, there's not going to be any market. No one's going to have a market for heifers because everyone is going to have more than enough.*

We just end up shooting ourselves in the foot all the time. We get better and better at what we do and get penalized for it.

Diversification has always been a way for businesses in general to survive in our rapidly changing economy, and each farmer in town has thought about its implications in the few off-moments they may have available to them during the course of the day. As Patricia points out:

Five generations ago it wasn't a dairy farm, so maybe with the way the economy is it won't be a dairy farm. Maybe it will be a goat farm. Maybe it will be a hay farm. I don't know. Hopefully it will be a farm. Dairy cows? I

Plate 40: Clouds covering the fields and pond at Cool Breeze.

don't know. It's an awful lot of work for what you're getting back for it. Hopefully it will be a farm. Years ago it used to be all apple orchards. All these fields have names – the East Orchard, the West Orchard, the Old Orchard. It was all apples – there weren't any cows out in those fields.

One route all three women on the farm have considered is producing goat milk. Patricia goes on to say:

Goat milk might be a good alternative. I think the way people are focusing on nutrition now,

I think there is a movement away from cow's milk. Goat's milk is perceived as healthier. There are actually more people in the world that drink goat's milk than drink cow's milk. We're the only country where there is a higher consumption of cow's milk. Goat's milk is what ninety percent of the world's population drinks. It's easier to digest. So I think there's a huge potential market out there – more for goat's milk, especially if you eliminate the middle man. The Farmer's Cow – that organization – has eliminated the middleman. They've taken that step. The way we in town here are doing it, it's the middleman that's making out on our efforts to produce milk – none of us are making out. If you go with goat's

milk and eliminate the middleman, I think the potential is there to be much more profitable. Just seeing the trends, the nutritional trends that are out there, especially if it's grass-fed, organic. If you can get those labels on it, that's what people want. They don't want the commercial.

Esther has also considered the switch to other animals in the future, but realizes there are problems to be overcome:

Goats are looking pretty good to us now – if we could get rid of the milk. I don't know how you are going to get rid of the milk here. No one picks up the milk. You'd have to probably hire somebody to take it somewhere. Of course the first thing you've got to do is get a license, which is plain stupid. If you have a cow license you ought to be able to milk goats. Same with making the cheese. You're not supposed to make it out of raw milk unless you have a license and all this. There is a big call for it, but we just don't have the time to investigate how to get rid of it. But someday we will be strictly goats. There is a market for kids – she's already had three, four people asking her for kids. Goat kids are going – boy, they can't get enough of them – and they bring more money than a calf even. I used to have rabbits. I think I've got to go back into rabbits. We saw one this week in the market bulletin for sixty dollars. I hope it wasn't one they were going to eat.

Linda milks her goats and makes cheese for home consumption and has sold a number of goats to others looking for them. After working with cows all day, the goats seem a nice break:

My sister had goats and we wanted a couple of places cleaned off, so we got some goats, and four became ten.... I've been playing around, milking the goats and making cheese. So now I'm down to seven goats that are all bred, that are going to be kidding. I had eight last year. I sell the kids and make a little money on them, but I think they're more of an expense. They're cute, they're personable, they're easier to work with than cows. There are several farms that have switched over to goats – I've heard of a couple of them up in

Plate 41: Linda milks one of her goats. Perhaps someday Cool Breeze will switch from cows to goats.

Plate 42: Goats occupy the high ground in their pen.

Granby and Lebanon that have fairly large herds. The price of goat milk is quite expensive, but then you have to find a market – it's not like cow's milk. I give them credit, but it would be a lot more work, because you're doing your own marketing, you're production, and adding value to your product.

Connecticut is no longer the best place, perhaps, for the dairy industry to thrive, and the state now has less than one hundred and fifty dairy farms operating within its borders. As Patricia laments:

This isn't farm country anymore. This is suburbia. Your land is worth too much for somebody to put a house on to be growing corn or grass in this part of the country, and it's too bad, because once you put that house up you've lost that potential to produce that food. You never get it back. At some point, food in this country won't be quite so cheap as it is now, and then all of a sudden farmers will be looked up to instead of looked down on, but we haven't reached that point yet. There's a lot of potential for ways to farm and make a profit, but you can't do it the way you did it ten years ago or even five years ago. You've got to look at what the consumer wants and make that change. Whether it's going to grass-fed cows, where you're marketing grass-fed milk, or making the cheese out of that milk and selling it as a farm cheese. But to just milk cows, send it on a milk truck to Garelick's or wherever, I don't think you're ever going to make money like you could years ago doing it that way.

In order to survive in the long term, Patricia sees that many changes may have to occur on the smallest of the dairy farms in the state:

You've got to eliminate that middleman and catch that profit for yourself that other people are making off of you right now. If you've got the ambition and you want to do that, then I'm certain you can probably make a good living at it. We've got the population around us. If we were out in the middle of New York state, I don't think you could do it, because there's nobody out there to buy your product. We're in the ideal situation right here with the amount of people around us. It's a big advantage if you want to take advantage of it. Like Buttons – they got out of the cow-milking business. They sell ice cream, they sell beef, Christmas trees. You know they're making more money than they ever did milking cows. If you want to do that, the potential is there. People are hungry for that experience. I see it here just with the goats. You cannot believe the number of people that stop. Haven't you people seen a goat before? And they are just so ecstatic if you pick up a kid and put it in their arms. I'm like – it's a goat. But they've never ever had that experience.

When asked how she views the outlook for dairy farming in Connecticut in general, Linda's response was somewhat tentative:

Plate 43: Dumping chopped hay into the truck.

It all depends on how the economy turns around. Connecticut fared pretty well so far, but I don't know. I don't know how much farmers can lose and still keep going. I've heard of farms that can't afford to go out – they owe so much money. The economy has hurt the dairy industry quite a bit. I think the state helped a lot this year, but how is the state going to keep bailing us out? It's not the answer.

I think the way the milk is priced for us needs to change. Maybe the price needs to go up in the store, but it kills you when you get ninety cents for that gallon of milk and the store is making at least that much and all they're doing is displaying it. Things are so nationalized. California has more milk so we get paid less. There is no surplus of milk in New England – there's a shortage if anything. But I guess they can bring it in from California or Wisconsin. If farmers got paid what it cost them to make the milk, and maybe a little bit more, I think they could survive. When you lose money year after year after year… . I never thought I'd see it as bad as it was this past year. The price of milk hasn't differed much since I was born. I think the farmers need to be paid what it costs them to make it.

The four last family dairy farms in town have one thing in common, unlike the newer and often larger farms that started up recently, and that, Esther believes, is one reason for their continuing operation:

Plate 45: Patricia adjusts the round hay baler before use.

These are old-time farms and had money behind them when they took them over. All these farms have been left to generations down, and a lot of those farms have not. And, a lot of those farms wanted to get big, big, big in a hurry and they borrowed a lot of money to get started up, and then what happened? Milk prices dropped and they're sitting there holding all this borrowed money. The farms that are in this town, except for those new boys that are just trying to start up, have been farms for years, and years, and years, and it makes a difference if it's just left to you or if you've got to go out and purchase it. So the farmers in this town are quite lucky to be in the position that they are. But you can still get in over your head if you don't watch out. I'm noted for being very tight, but we're still here. If you don't have money behind you, it gets pretty rough. And then too, it is also

the way that you live and the way that you run things. If you go out and try to buy everything that there is on the market you can get yourself in trouble in a very short time. If you try to live within your means and make due a little bit, it makes a difference.

Despite the often gloomy outlook for the dairy industry in New England, the local farmer's natural optimism keeps them going. While Esther initially jokes when asked if she has thought about another line of work, it is easy to see her heart and soul really are in what she does:

Oh, many times! Many times! No, I like it. I joke that I don't,

Plate 44: Ears of corn ready for harvesting.

but I do. I worked different places before. I always liked work. Work doesn't bother me – I don't care what I do. I would just as soon pick stones or milk cows or whatever – it doesn't bother me what I do, but I do think that this world is upside down when you have to work ten, twelve, fifteen hours a day and then you hardly make a living. We used to get paid for milk enough so that you could make a half-way decent living, but now it's completely ridiculous. Of course milk prices are up a little bit now, but for how long? Next thing you know they'll be down and there's nothing you can do about it.

Patricia's reaction is much the same, and her feelings seem universal to every farmer in town:

I've thought about it, but I can't think of what else I'd rather do. Sure, there's days when you're just so tired of it and you think, "Why am I doing this?" And then I think, there really isn't anything else I'd rather do. When you don't get enough sleep – trying to do all the work and take advantage when you get really good weather and get as much fieldwork done as you can – you get short of sleep sometimes and it does get to you after a while. On days when the weather is nasty, you think, "What am I doing out here?" But then you get days when the weather is really nice and you see all these school teachers going by, going to work, and none of them look too happy, and I'm just sitting on a tractor smiling away. It's like anything – there's advantages and disadvantages and you have to decide if the advantages outweigh the disadvantages.

Plate 47: An unused disk harrow sits alongside a stone wall.

Plate 46: A cow waits to leave the milking parlor.

Plate 48: Rare photo of John and Patricia Palmer together.

**Scenes from
Cool Breeze Farm**

Beriah Lewis Farm

The Beriah Lewis Farm holds the honor of being the oldest continuously operated family farm in North Stonington. Purchased in 1791 by Beriah Lewis from the estate of a local man executed for thievery, the farm is now in its eighth generation. It is currently owned and operated by Rosalind Lewis and two of her sons, Ted and Ledyard Lewis. (One of Rosalind's daughters and son-in-law also own a large dairy farm in Lebanon, Connecticut.) Rosalind started out as a school teacher, but she eventually married Dave Lewis and moved onto the farm. She continued to substitute teach until the birth of their fourth child. As she did not come from a farming background, the experience was all new to her:

It was very, very interesting to me, because I knew very little about farming. I never liked animals – never. I don't like them today. I'm not afraid of them, but they're not my expertise I guess you'd say. So it was interesting to me. And it was interesting to me to learn nutrition, and that animals were just like people. They have to be taken care of, and their health is important, their feet are important. To get milk they have to be comfortable. There are a lot of lessons to learn. My friends that I went to school with and

Plate 49: View of the farm from the cornfield across the street. The grass cover crop prevents erosion.

taught school with are appalled that I am still with it. However, they come here and find it very interesting. I look at my family and I think my family is very blessed to have been brought up on the farm – the values, the work.

Even though she never worked with the animals, there was plenty of other work for her to do in the daily operation of the farm. Beyond raising their five children, she goes on to talk about her other chores:

I did all the books, cooked, and kept them all going, and that's a full time job. I used to feed the help – the help would come in to eat. We didn't have a McDonald's or a Tim Horten's that they could run to. Particularly during harvest time, they would come in and eat. Ham, baked beans, macaroni and cheese, stew. At harvest time they used to come in in shifts – you know you had to keep the trucks going. Probably at that time there would be eight or ten in my family. The books are a full-time job, even fuller now – milk inspectors, diesel, gas, registrations, OSHA, and insurance.

The farm was smaller when Rosalind and Dave married, although perhaps typical in size for farms in the town in the mid-1950's. Dave was still using horses at the time:

When we were first married, they had work horses here. The only time I remember oxen was as a hobby. There was a horse barn right out there where that shop is, and I think there were two – two teams. When I came here we had forty cows in 1956. When my husband died in 1997, we had 600 cows, and today we have over a thousand. Calves, dry cows, milking cows, and beef cattle. Dave's grandfather lived here. He was ninety-eight when we were married, and he died when he was one hundred. After he died, Dave inherited the farm. We fixed just the downstairs up, and moved here with one child. The other children were born here, but my oldest was not. My boys are the first generation to have more than one boy, and we had three. Previously, every generation had, well, they had more than one boy, but they didn't live. One boy always lived and became instrumental on the farm since 1791. And my boys, there were three of them, and they all stayed on the farm. I often said their father was a wonderful mentor, because his three boys wanted what he wanted for them.

Although the cows were no longer milked by hand at the time Rosalind came to the farm, it was still a more difficult task than it is today with the modern milking parlor:

Plate 50: Wind vane on barn roof.

They had milking machines when I came. They had two machines, but they carried the milk – the milk tank was here – they carried the milk the length of the barn and dumped it into the tank. They called them milking stations, on wheels, with four milkers and two cans that you'd fill, and then they'd put them in the tank, and the tank was chilled. I think I remember that cold water chilled it. They came and picked it up by truck. It seems a long time ago.

Transitions in the dairy side of the business were not, of course, the only changes to be made on the farm. Rosalind describes what life was like in the household end of the business as well:

When I first came here we had ducks and chickens and sheep, and they would run around in the yard, and they'd dress them and bring them in and I would have to cook them, and I

Plate 51: Rosalind Lewis in her kitchen.

had to learn. My mother-in-law used to take a hen that had laid and was all done, and of course it would be dressed, and she would put it on the back of the wood stove in a pot, and it would be there all day, simmering, you know, and then she'd take it out, and she'd cut it up, save the broth, and she'd roll it in johnnycake meal and cook it in bacon fat – it would just melt in your mouth. Oh, it was delicious! Then she'd take the broth and make the best chicken soup you ever put in your mouth! When I came there was no plumbing. This was three rooms. The wood stove was there, the sink with the pump was here, the flour barrels, and the johnnycake meal barrels and the sugar barrels were here. And this was the potty chair. Grandpa was one hundred,

Plate 52: Spring mud charts the paths of tractors and the equipment they tow across the barnyard.

They had to do chores. They started at six or seven. They had to get wood for their grandmother, they had to feed calves, they helped their father haying. They were brought up in it. The boys were all athletes, and they had to do chores just the same. I remember Ted and Hummer, they would have a baseball game, and when they got through it they would have to come home and milk or hay. They couldn't just hang out. There wasn't much wasted time. Oh, they played hard, but they didn't have a lot of time to waste. The oldest daughter would love to be on the farm. The youngest daughter did what she was told, but she wasn't quite as into it as the oldest girl.

While their children never really took to gardening, Rosalind talks about how they taught their children the value of work and the benefits it could bring them:

They used to plant sweet corn with the corn in the fields. The sweet corn came off before the field corn, so they planted it in the paths where the trucks would go and along the walls where you don't plant because you can't get in to cut it. The kids would pick it and we'd take it to Westerly or sell it here at the house, and the corn money was theirs – theirs to decide what we were going to do with it.

Some of the money they earned would go into the family's yearly summer vacation. While Rosalind usually decided where they would go and what they would do, there was some early dissent among the ranks:

The oldest girl didn't think that I was doing a very good job at deciding. She was in sixth grade, and she came home one day and she said, "I know where we're going this year." And I said, "Where?" She said, "I've got just the place. We're studying about the Amish," and she told her father. Well, a farmer down here in Westerly knew some farm down in Pennsylvania that had purebreds – Aldo – and he got Dave interested in the farm with purebreds. Well, we made the reservations – I will never forget

and there weren't many changes until he died. Dave and I did the downstairs when we moved in – just the downstairs. When each child was born, we'd do one more room upstairs. There's no heat upstairs even now. Out there was an outhouse. Out here there was a little house with a big pit underneath, and they would hang the hams and butter. It would be cold, you know. And where that windmill is, that was a cistern, and all the water from the house ran into that, and that's what they used to wash clothes, in a wringer washing machine.

Dave and Rosalind's children, growing up right in the middle of the farm, had work to do beyond the normal activities of going to school. She relates:

it – it was at the Landis Valley Family Motel. It had a restaurant and it had bunk beds. So we packed four children. Ted and his brother were in the bunk beds, and Myra and her sister were in one bed, and Dad and I were in the other. We went to the restaurant the next morning, and they had oatmeal, and boy, everybody thought that was the place! And that was many years ago and I guess we've been going to Pennsylvania ever since. We have friends there. The Amish come here and we go there.

After finishing high school, all three of the boys decided they wanted to stay and work on the farm where they were raised. This meant changes had to be made, as Rosalind explains:

If you were going to stay in business, you had to make a living, and in order to make a living you had to get bigger. The farm expanded with each generation. Each generation added acreage, and with the acreage both cows and machinery. Because we were only milking, say, a hundred cows, and now we were talking about three families – Dad's family, and the boys were now out of high school and looking forward to families, so now you had not one family, but you had to make a livelihood for three families.

And so the three sons began to work the farm with Dave and Rosalind. Hummer Lewis, their third child, was instrumental in running the farm with his two brothers until his death in 2005. Rosalind recalls that difficult time:

Plate 54: Ledyard Lewis – a seventh-generation Lewis.

Plate 53: A calf looks out on the day.

When we lost Hummer, it was very, very difficult, because now I have the oldest and the youngest boy, and that was a tough period in our lives, but we got over it as best we could. It crops up once and a while, but basically, the boys are doing very well. They work together very well. They know their strengths and weaknesses, and they accept them and accept one another. If one can't do it, the other will pick up the slack.

Ledyard recalls the skill and knowledge his brother Hummer brought to their team:

Five years ago we lost a brother, and he was definitely in charge of the cows. He could do things a little bit better than my brother and me, health-wise, on the cows. He knew all the cows. He could tell you a cow that was ten miles away from here. She didn't even have a number in her ear, but he had a number for her, and he could tell you to go get 639, and she had trouble calving last year, and she hangs up a little bit, and her right hind toe is a little long.

While their mother keeps track of the vast amounts of paperwork involved in running the business, Ted and Ledyard oversee everything else outside the farmhouse walls. While each must occasionally work on all aspects of the farm, they tend to have areas where most of their work is concentrated. As Ted describes:

We pretty much run it together. When Hummer was here, he used to do a lot of dealing, you know, in cattle. He didn't have hands-on on the farm as much. After Hummer, now Ledyard and I kind of take care of the animals, and the help, they do the fieldwork and the feeding. That's the way it works. It would be nice if I could get out in the fields more, but with the situation, it's best the way things are going. My father used to say you never saw a real good farmer at both. Usually they really excel at the cows, but it's pretty hard to get one to do a real good job at both. With a lot of brothers, one does the crops, one does the cows. But there's not too many farmers left now, so you have to be pretty good at what you do.

Ledyard continues to describe the feeling of working together with his brother and his mother:

Plate 55: Liquid manure is sprayed onto the fields in the spring before the field is harrowed.

The skeptics were proven wrong, however, and the farm continued to expand. Ledyard continues:

Now we milk 310 cows. We like to milk 330, 340 in the winter months. In the summer months, the cows don't give as much milk. I would like to milk a few more cows now. In the summer, some cows we dry off and we put 'em out on grass, especially if they're going to have calves in the early fall. In July and August there we don't make as much milk because of the heat. I would say we probably have, with the milking cows, three hundred and fifty-five total. Sometimes it gets up to three hundred and seventy, three hundred and eighty. That's with dry cows that are going to have a calf within the next two months and milking cows. There's a thousand head here total.

My mother is great with the books, here, and my brother and I are on top of the cows. I know that there's a stone or two that doesn't get turned over, but there's a lot of stones here that get turned over quite regularly. We work awful good together. I'm not saying that he can't run this place without me, and I know that I can run this place without him, but between the two of us here, we kind of dot our "i"s and cross our "t"s, and I think we make a pretty good pair. We went through a lot of adversity. Thirteen years ago we lost our father, and everybody said, "Oh, the Lewises won't go, there won't be a farm there anymore."

As the farm grew, it eventually became too large for family members to run on their own. Ledyard describes the types of work they divide between them and the positions left to men they hire:

Ted is more into reproduction and cow health, and I'm more into ordering, fieldwork, or seeing that tractors get fixed. We have two guys that are employed by us. One's in charge of planting and all of the corn, and one's in charge of all the hay. We oversee them, but they're kind of in charge.

Plate 56: Stainless-steel milk cans used in the parlor.

As with all farms, the day begins early and ends late. Ted describes part of a typical day:

I get here at 5:30 and I milk. We get done milking around 9:30, and I breed the cows if they need breeding. I breed them as I milk, and then we'll go up and clean the barn, take care of the heifers, feed the heifers, and then go over to Anthony's and see if there is anybody ready to calve. You know we probably check them two or three times a week – we have to go over there for an hour or so. We try to bring them home a month before they calve. In the afternoon we start milking at 1:30 and we get done around 6:30. We then milk at 9:30 – we have hired help do that. We're milking three hundred and thirty cows at the moment. We've been over three hundred for the last twenty years. We built this milking parlor twenty years ago, twenty-one years ago. It holds twenty-two cows, eleven on a side.

As Ledyard has two young children at home, he comes in a little later:

My brother gets up earlier than I do in the morning. He's here at 5:30. He goes home and takes a nap in the afternoon for a couple of hours. Then I get here between seven and eight o'clock and I work until eight o'clock at night, seven days a week most of the time. In the winter, we don't usually take too many days off. We try to take one week out of the year that both of us can go and kick our feet up and not worry about cows. I usually work like twelve to fourteen hours a day, and he's about the same seven days a week. In the summertime there could be days that we put in hay until ten or eleven o'clock at night. When we cover the corn pile with plastic and tires – we try to put one piece on a night – after seven or eight o'clock at night, so usually we don't get home until nine or ten. This year we were kind of lackadaisical and we didn't get it covered so we spent a whole Saturday and we covered the whole pile from nine o'clock in the morning to five o'clock at night. We threw a lot of tires.

He milks the cows in the morning. He's in the parlor all the time in the morning, from 5:30 to like ten, eleven. I'm in the parlor from like 1:30 in the afternoon until seven at night. I see him when I get here first thing, and he usually tells me to get out of the parlor to do something. If he needs a hand in the parlor, I'll give him a hand, but if he tells me to go get a load of sawdust or call this guy and get this tractor delivered, I'll do that. I'm not usually involved too much in the parlor in the morning. Then sometime between eleven and 1:30 we have a chat and he tells me how things went in the morning, and then he comes back from his nap at four, four o'clock in the afternoon, and I tell him what I did so far, and he usually goes outside and I'm in the parlor. Our feed man takes off one week out of the year, and then we have to feed the cows. Normally the cows are always fed – we don't have to worry about that. He plants all of the corn too – he's in charge of the corn, so when we're choppin' corn we have to feed the cows too.

Unlike the other three farms in town, Beriah Lewis Farm milks their cows three times a day. Ted explains the benefits to this extra milking:

Plate 57: Silos for grains brought to the farm by truck.

Plate 58: View through loft door into barnyard.

There is less stress on the cow, and a cow milking a hundred pounds will milk a hundred and twenty. I don't know if there's twenty percent more, fifteen to twenty percent more production you can get, depending on the lactation of the cow – if she's fresh, if she's peaking. If she's giving a hundred and twenty she's going to milk a hundred and fifty three times a day. When we went to milking three times a day it was also good for our families because we'd be done at six at night. If we milked twice a day we'd probably be done around eight, so that was a good thing when the kids were younger. Now we could milk at three in the morning and three in the afternoon and we'd probably still get done at eight. It has its pros and cons. Three times a day you've got a little more room for error. Like the late shift, if they don't do a good job, you can catch that in the morning, and the cows aren't as full, so there's a little more room for error in the milking part of it.

As with all modern high-production dairy farms, the nutritional requirements of the herd is carefully monitored. The farm works closely with a nutritionist from the company that supplies their grain. Of course, there is some incentive for the nutritionist to push grain on the animals, although every farmer keeps a close watch to make sure the animals are getting what they should be getting after factoring in all of the variables. As Ledyard discusses:

I talked to the guy last fall, to our nutritionist, and just like anybody that you do business with they always look out for themselves, and he showed me a chart that we were feeding forty-five percent corn silage and hay and fifty-five percent grain. And I told him, "We have plenty of corn silage and we have plenty of hay. I don't want to give all my money to you. Let's change that percentage." The last I knew, a couple of weeks ago, we were feeding fifty-one percent corn silage and hay and forty-nine percent grain. For what we get paid for our milk to what the grain is, I think that six percent would be well worth the other way.

The milk-producing animals at Beriah Lewis Farm are also divided by age into three main groups for feeding – calves, heifers, and milking cows. In addition, the farm has a herd of one hundred and fifty beef cattle that receive a slightly different ration from the milking cows. They eat a grain designed for them, hay, and corn silage. Much of the cost of feeding dairy cows in New England comes from the grains they must import into the area. The primary reason dairy farms in western parts of the country can get so large on relatively small parcels of land is because the grains used to feed them are grown locally. As farms disappear in this area, major suppliers begin to close down infrastructure once used to support them. Ledyard discusses the implications of these closures:

Plate 59: Ted Lewis – another of the seventh generation.

All of our grain comes from Cargill. They had a plant in North Franklin, Connecticut, which is forty-five minutes from here, and they said that with the number of farms around that they couldn't continue to operate, so they shut that one down. The main plant is in Salem, New York, and that's four hours away to get here, so they truck it quite a ways. We take a whole tractor-trailer load at a time. There is twenty-four ton on a tractor-trailer load. That's just three or four different types of grain, I think that lasts us about six days. Another thing that goes into the cow's grain is brewer's grain – it's hops from beer, and we get that relatively cheap. That comes from Anheuser-Busch in Newark, New Jersey. We go through a trailer load of that every six days. And then we have beet pulp. We go through a trailer load of that once a month. We buy twelve trailer loads of that a year. It doesn't come from the US. They ship it in by barge through the Saint Lawrence Seaway, and they truck it from Ogdensburg, New York. We use molasses to help the cows. We get that by tractor-trailer load too, and we get a load of that, probably, every four or five months – twenty-four tons of that.

With a thousand head of cattle located on and around the farm, preparing feed and delivering it is a serious affair. The ingredients for each group of animals are blended together in a mixer wagon in nine thousand-pound loads. Ted and Ledyard have had John Clegg on their staff to handle this enormous task. John's background, as with many people in the business today, began with family members in the business:

My family has a farm in Franklin. My grandfather started a farm up there – actually they're still in business. It was in the early fifties. He's not around anymore, and my uncle and my cousins are running that. My father had somewhat of a farming background – he was from this area too. He was involved with 4-H and things like that when he was younger. He became a professor in an animal science department. That was out at Purdue, so I lived out in Indiana for eight, nine years and then we ended up back here. I worked at the home farm for a number of years, and then I guess when I was twenty-four I left there. I worked for a guy in Preston – purebred animals – and I did a little bit of everything there. He sold his cows and took a year or so off from milking cows, and in the meantime I ended up coming here. That's all I've been doing, ever since I was a kid. I always had a love for it. I used to come out here for whole summers when I lived in Indiana. You know, stay with my grandparents and work on the farm there all summer until I had to go back to play football.

John's family also has had a long association with the Beriah Lewis Farm:

I've known them a long time. I knew them when I worked at my grandfather's – they were friends. Their father used to come up and pick up our cull cows and so forth. Take them to the auction for us and help us different ways – we helped each other. I've known them quite a long time, probably around twenty-five years now I guess. I've been working here for thirteen.

Plate 60: Manure spreader used to distribute solid manures collected on the farm.

Although John was the herdsman at his grandfather's farm and really considers himself a cow person, his role working for Ted and Ledyard is slightly different. While he occasionally works with the animals, most of his time is spent outside of the milking parlor. He goes on to talk about one of his primary areas of responsibility on the Beriah Lewis Farm:

I do all the mix-wagon work. Somebody else feeds the baby calves, and somebody else feeds the calves up in the barn there and the heifers that are out in the pastures. I'll load the feeders a lot of times because I'm out there feeding, but somebody else usually does that. I just feed the milking cows, the dry cows, and like I say, that transition, the pre-fresh group, and one group of heifers. The reason we don't feed all of them with the feed wagon is that it's just not practical. Either they're a long ways away or there isn't the right type of bunk. I do, for the most part, just the mix-wagon work. I feed the beef cows with the mix-wagon in the barn over there, but there's quite a few of them, so they get fed outside too. They don't get fed that much different then the older heifers, but I think Ledge gives them some other stuff over there – a little bit – not too much. They mostly get hay, corn silage, and brewer's grain. Brewer's grain is just a cheap source of protein and it seems to work real good with the corn silage.

He also notes the importance of the nutritionist and the testing of the feed going into the mixer every week:

We have a nutrition specialist that comes every week, once a week. He'll walk through the cows and most weeks we don't change anything, we don't do anything different, but just try and keep on top of it, because things change. Ingredients that I'm putting in can be different. The corn can be different, even through the course of a year. Different fields will test different. It's a big enough pile so you actually have layers – some of it's different varieties too. We test the silage for all the nutrients – hay too. We test all that. He does all the balances. You've got to work with him though, because what works on paper doesn't necessarily work practically. It's another thing you have to try to balance – practical to paper. He works for Cargill. That's who we've been working with for quite a while now. They're pretty good – I mean that's their job, to work with the farmers. They're not just salesmen; that just doesn't work. They'll get tossed and you'll get grain from somewhere else. They want to sell grain, but they don't go overboard. And every farm is different too, you know. Every farm is different.

John describes how the animals are grouped for feeding:

The animals are grouped according to age and size – more or less age, but if there are some of them that

Plate 61: Corby Steadman spreads clean sawdust in the calves' stall in the old stanchion barn.

Plate 62: The road through Beriah Lewis Farm.

are way bigger then they'll get bumped up to the next group, and they'll be younger than most in that group. For the milking-cow ration, the way they do it here is that all the milkers get the same ration. Even though they're in three different groups, they all get fed the same thing, which in my opinion is the way to do it. The way it is here, generally the older cows or any cows that are injured or don't like to use the stalls in the new barn end up in the old barn, which has sand in the stalls. It's better on their legs. Because we don't have different rations for different groups, we don't divide them by production. That's how you would do it if you had a different ration for each group – you'd have different production groups. Here it's where the cows will

do OK. Generally speaking, the younger ones and all the healthy cows end up in the new barn with the mattresses. And if at some point they're not doing well, we'll put them in the sand stalls.

John does all of the mixer-wagon work. For a measuring cup, he uses a payloader:

First thing I put in will be a commercial grain mix, which is roughly twenty-five percent protein. It's about 20 pounds per animal per day on that. Then beet pulp. I'll feed them about three and a half pounds per day per cow. Hay varies a little bit, but now I'm feeding a little under three pounds of hay per cow – dry hay. Some of it first-cut, some of it better. It depends on what we have available. We're at the end of our year now, so we run out of a few kinds of hay, but we try and feed the cows real good hay, preferably alfalfa for the most part. Fifteen pounds of wet brewer's grain, which actually comes from the Budweiser brewery in New Jersey. That's wet. It's fifteen pounds, and it sounds like a lot, but it's wet. We get a tractor-trailer load about once a week – twenty-three or four tons. We use a little bit of that for the heifers too, with corn silage. And for probably the last eight years, we've been feeding about a pound and a half of molasses – just straight cane molasses per cow per day. Right now about eighty pounds of corn silage per day. They're gluttons. That's what they do.

Each of the milking cows is getting just under one hundred and fifty pounds of food per day. This twenty-five tons of feed each day is just for the cows that are actually being milked – roughly one third of the total number of animals on the farm. Expenses can add up quickly when just the grain fed to them costs the farm over a thousand dollars a day. Of course the calves and heifers are not fed nearly as much. John continues:

The baby calves get milk, and they'll have a little bit of what they call sweet feed, which is either a sixteen percent or twenty percent pellet with quite a bit of molasses in it. They'll have that available to them to eat as soon as they'll start eating it. The sooner they'll eat it the better, and a little bit of hay available. I always thought the faster they get eating grain and hay the less trouble you'll have with them, as far as get-

Plate 63: A pair of Ted's oxen in an open-air shed.

Plate 64: Ledyard and crew give an IV to a cow that just gave birth. This helps replenish the minerals lost through the birthing process.

ting scours and stomach problems. And after that they'll get moved into a group pen, because the little calves are in the hutches individually. They'll get put into a group pen for a while, and then they get weaned. By then they'll be eating a decent amount of the sweet feed and the hay, so you just give them as much of it – not as much grain as they want, but you want them to have as much hay as they want. As they get older, you kind of ration them down as to how much you have to add.

Once they get up into the barn they'll start getting as much corn silage as they want along with hay, and they'll get a sixteen percent dry pellet. As far as the rate that they feed, you always figure a pail for ten calves. A five-gallon pail for ten calves twice a day. As they get older you don't have to give them a special grain. They eat enough of the other as long as you have quality forage – they do good on that. They don't need much else.

Next they get over with the bull, and when they come back here, a couple of weeks before they calve, they'll go onto a pre-fresh ration, which is light. You don't want to push them like the milking cows, because they're making up an udder and stuff, and it will actually cause you problems. They'll udder up too much – it actually wrecks their udders if you give them too much grain. You can also run into a lot of health problems that way. But, if you don't do anything, you can run into a lot of health troubles also. We try and get them for at least two weeks before they calve. It's more like a transition to the milking ration. It's a pretty good shock to their system to go from eating hay and corn with just a little bit of brewer's grain and then get thrown right into where they're getting twenty pounds of the high-test and all the other stuff. That's the main thing there – I think that helps. You don't always get them home when you want them, but that's what you try to do. You try to get them out here, wherever they're going to be for a couple of weeks, and I'll mix that up separate. And they get about as much hay as they want until they calve. And as long as everything goes well, you put them right in with the rest of them.

Without calving, cows do not produce milk, and without calving occurring all through the year, the milk production on the farm would cycle up and down. For that reason, cows are bred throughout the year. According to Ledyard:

The professors say that you want them to calve out at two years old, but a lot of farmers are with us and agree that's a little bit young. We keep 'em twenty-six, twenty-eight months, you know, just over two years old, until they calve out. They're just not developed, and then their first calf, if they have a big calf, it comes out hard and they're not developed enough. Then, when they start milking, they don't come into their potential – I mean it takes them a whole year. You can almost see 'em grow when you're

Plate 65: Martin Rodriguez with a new calf.

trying to get the milk out of them. I had a cow I bought off a guy and she calved at a year and eleven months, and another farmer came and saw her when I bought her and said, "Boy – she's kinda small." And I said "Yeah." And I sold her for hamburg like five years later and she weighed 1900 pounds. That's a prime example. I saw her grow while we were milking her.

They should be bred every sixty to eighty days. That's a whole other diversion from milking – getting them bred. We push our cows so much that we have a tough time getting them bred back to have a calf. If they don't get bred, with the feed that's going in them, they start getting fat, and when they get fat, it's tougher to get them bred. We have cows here that are twelve, fourteen years old. They should calve once a year. I think I saw somewhere on a piece of paper that our calving interval is fifteen months. The ideal is you'd like to get that down to twelve.

Beriah Lewis Farm has been taking advantage of a program sponsored by Tufts University to oversee their breeding program. Ted explains how this works on the farm:

Tufts comes every two weeks. They do health checks and pregnancy checks and reproduction. Besides that, we do most everything else here.

Plate 66: Elizabeth Lewis loading hay bales into the mow in the barn.

We do all of our own insemination. We use a bull in the winter and this time of the year we artificially breed up pretty near everything. All our cows – the heifers – we breed with little bulls. About a hundred percent artificial breeding at this time of the year. In the winter we use a bull. In the spring, until summer when it gets hot, we use a bull again, and then in the fall it seems to be harder to get the cows to stay settled and get bred when it's real hot out, so we use a bull because we're also going to fairs, and it's just more convenient for us. Then in the fall I'll breed artificially pretty much, and then January we use a bull for a couple of months to clean up the cows that didn't get bred in the fall.

We're not into genetics as much as most farms. We buy good blooded bulls, and as far as semen, we buy bulls that are good on feet and legs – we stress that. We seem to be able to get the milk out of the cows – we can make them produce. In a free-stall barn we stress the feet and legs – that's what we look for in genetics. That's the first thing that goes on them. They're pushed. Our cows average eighty-five, ninety pounds a day. If they averaged sixty pounds and they went out on pasture during the day, you probably wouldn't stress feet and legs, but you know on cement 24/7 it's a lot of stress. Our herd is primarily Holsteins. We have some Jersey-Holsteins, but mainly Holsteins. Ninety percent Holsteins.

Tuft's sends both professors and students in their veterinary program to working farms so the students can receive hands-on experience in the field, so to speak. Learning outside of the classroom in real-world conditions helps prepare them better for entrance into the working world, and acquaints them with a variety of farms as well. In the past, farmers would often have to bring sick animals to the universities' animal hospital for diagnosis; the advantages of having vets come to the patients are obvious.

Ledyard explains that many factors go into how much milk a cow can produce and how that has changed over time:

We have one of the best production herds in the state. We have cows that have just calved, and are going along, that are giving a hundred and fifty pounds of milk a day, and then we have cows that we cannot get bred that are on their way to going to hamburger that are giving fifty, sixty pounds a day. We try to average around eighty-five to ninety pounds per cow a day. It takes a lot of time – my brother and I spend a lot of time with 'em. Forty years ago, they would get forty pounds, but then they only milked them twice a day, and they didn't have the grain trucks coming in. You gave them a scoop of grain on top of their silage at night and you got the milk that you got out of 'em, you know?

The other major job on every farm beyond working with and tending to the cows is producing the feed to keep them fat and happy over the course of the year. While New England farmers must import their grains from the West, they still produce most of the mass of feed that the animals receive on a daily basis. While their cows eat some twenty pounds of grain a day, they are also consuming one hundred and thirty pounds of silage and grasses that are grown on the fields the farm owns or rents. As with many other tasks farmers have done throughout time, technological changes have increased their efficiency in the fieldwork

Plate 67: After milking, the cows will head back to their stalls on their own.

they pursue. In the past, dairymen had to cut the grasses and give them enough days to thoroughly dry before they could bale and store them in barns, as improperly cured hay could become moldy, or even worse, spontaneously ignite, burning down their barns. While they still use dry, baled hay, newer technologies and methods have reduced the time required to get all of

Plate 68: A pig rests comfortably in its pen.

their crops under storage. Ledyard explains one new baling technology in use:

We used to bale all our hay dry, up 'til five years ago. We bought a round baler five years ago, and now we bale wet hay in wrap. The main reason behind that is we do so much land that we couldn't get it off by July – we were still doing hay in August. We could only go so fast. With the wrapped hay, you mow it and you let it dry down like fifty percent, sixty percent – I don't know, there's a window there. Say fifty percent – about half dry, and then you bale it and you wrap it and you let it cook. It's almost the same thing as haylage, but haylage you mow it and then you chop it and that comes out finer. We don't have a chopper, and we don't do anything with haylage. To make dry hay you need four days – four to five days. Alfalfa, second-cutting alfalfa, if it's real green it takes five, six days. With wrapped hay you just need a two-day window. As far as the wet hay, you don't need a barn – you can wrap it and leave it outside. Also less time consuming in the spring – you can get it off the field and get to second cutting quicker.

While this saves on labor and allows them to get in all of the grasses they need, it does introduce new twists to the feeding of the animals. He goes on to explain:

We have so much moisture between the silage and the brewer's grain, and you don't want to get too much moisture, because you get a lot of acidosis, and that, with the milking cows, blows up their feet. You have to keep your nutrition right.

If we get feeding wet corn silage and wet brewer's grain we can see a problem with the cows. It's not just their feet, but it's their stomachs too. So we feed dry hay to our milking cows mixed in with the silage. Right after we chop our corn, it seems like October, November, it takes us a month or so to get the nutrition level. We always have trouble with feet in October, November, because the corn's cooked, but there is still a lot of moisture in it. It hasn't all drained out and you're feeding it to the cows. It will last two or three years, but that's another thing with corn. They say that the starches build up in it if it sits for a long time, and then you get the other side of the podium – that you have too much starch in your corn and you have to cut back on that. You've got to keep the nutrition level the same.

Like every other business facing tougher and tougher economic conditions, farmers must find ways in which technology can help lower overall costs. Traditionally, fields were plowed and harrowed, which is not only time-consuming, but hard on the equipment in gravelly soils. Depending on the weather, a field could require harrowing three times before planting. John Clegg notes that Beriah Lewis Farm is trying to eliminate this expense by spraying the cover crop with Round-Up this year in another cost-cutting move:

Because we spread so much manure and we run on the fields with equipment and get compaction and ruts sometimes when it's wet, we kind of have to harrow

it. Normally we'd harrow it early and then try to get manure on it and harrow that in. This year we're doing things a little different to try and save money on harrowing because that's gotten to be expensive to do too, between fuel and equipment. You wear out those harrow disks – one of them I think has like eighty disks on it, and they're like ninety dollars apiece. That's just for the disks.

Plate 69: Two wagon loads of dried hay ready to be loaded into the loft of the big barn.

The price of steel went crazy a couple of years ago. We're trying to avoid harrowing some, so what we did was spray the cover crop with Round-Up, and that eliminates at least one harrowing. You had to harrow three times if you just left the cover crop. You had to harrow the cover crop to keep that under control, and then you put manure on and harrow it again, and then hopefully you get close enough to planting time so you can just harrow it that last time, preferably the day before you plant it or the day you plant it. You don't want to let it sit too long, because you'll get weed pressure starting early before you plant. Some of the land further from the farm, we're actually no-tilling that. Some of it, if it's not too far away, we'll haul manure there, or we can buy chicken manure or commercial fertilizer and do it that way. That you don't have to harrow at all.

Many of the farms in town also employ no-till planting on much of their land, although it hasn't eliminated more traditional methods entirely. Much of the cropland in North Stonington, as the town's name suggests, is less than perfect.

As Ledyard explains, no-till planting can help with these conditions:

We just bought parts for our planter last year. We hired no-till done two years, and then last year we bought parts for our planter so we can plant. We have places that we can plant no-till. Our planter plants both. Most of the land around here we put so much manure on, and of course you can't always go out on a dry day and we always leave tire tracks – rutted up – so we try to harrow it and get it smooth. The no-till land we try not to go in there in the mud in the fall to chop it, and we try to keep the fields smooth so we don't have to go in and level 'em up. And most of those fields are rented fields. We don't know how long we're going to have them. We've spent a lot of time there pickin' stones already, and they're very stony. We have a hundred and thirty acres of land that we rent that is very stony, and that's why we plant it no-till. One field is fifty acres on top of a hill and that washes profusely if you do harrow it, and plus it is stony, so no-till seems to work pretty good there.

While all of the farms in town must buy some commercial fertilizers over the course of the year, they try to minimize this need where they can due to the enormous expense of these highly engineered products. Their primary source of nutrients for the soil comes from the manure their large herds produce. The Beriah Lewis Farm is no exception, and again, modern farming methodologies have not only reduced the labor needed to handle that material, but also increased the benefits of the manure itself. The large free-stall barn complex on the Lewis farm was built by Amish men from Pennsylvania, and introduced new methodologies for the handling of manure into the local farming community. Ledyard remembers when it was built:

In 1990 my father built the free-stall barn. Its got two hundred and sixty-four stalls, and its got nine-inch dams that hold the water underneath it. The manure goes through the slatted floors – it has waffle slats in it, and the cows walk on it and it goes through the slatted floors. It's all gravity-flowed out to this million-gallon manure pit that's out behind it. It's a lot less tractor intensive – you don't have to scrape the barn every day and it stays ninety-five percent as clean as if you scraped it every day. The only drawback I have against it is it seems that there's a lot more gasses in the air. Anything that's metal in there seems to rust a lot quicker. We've put two sets of free-stalls in there in twenty years.

While the manure pit holds a million gallons of manure, it is only coming from the two hundred and fifty or so milking animals in the free-stall barn. All of this manure is in liquid form, and therefore can be pumped and sprayed onto the fields, which is much quicker than spreading manure in a solid state. Ted lists some

Plate 70: Noah Lewis, one of the eighth generation, drives the feed-mixer wagon back for a refill.

Plate 71: An empty feeder is returned to the farm.

of the other benefits to their system:

The manure pit we put in paid for itself in two years. You put the liquid on and we don't have to buy fertilizer, where, when you spread every day like we were before, with the runoff, you still had to get fertilizer. You put it on, you disk it in and it stays there – you don't lose as much. Handling it is easier, and better for the environment too – it doesn't run off. We still use fertilizer on our corn and hay ground, but not as much. Close to the farm here we don't use any fertilizer, but further away we still have to buy it. It's just not efficient enough to justify hauling manure ten miles.

While all farmers have a time of year when they would prefer to spread manure to maximize its benefits, practical considerations can get in the way, as John describes;

In the early fall we'll go to spreading cow manure, mainly because we need to empty the manure pit to have room for the wintertime. You want it to be full in the spring, because you don't really end up getting much nitrogen value from the manure in the fall because there is so much nitrogen lost as it's spread right on the top. It's not tilled in or anything, but you do get a lot of other things that do stay right in the ground. Then we will do it again in the spring, as soon as the weather and the field conditions aren't too bad, and later, in March, usually, start over again.

Once the fields are fertilized and put into physical shape, the planting begins. Ledyard discusses their usual practice when it comes to timing:

We start in April and spread manure. After we spread the manure we harrow. We don't do any plowing, 'cause of the gravelly soil all over the farm. We plowed twenty years ago. From the gravel, the shears on the plow would only last an eight-hour day, and then you'd have to get new shears. We stay with harrowing, and harrow the rye, the winter rye underneath. A lot of farmers start the end of April, first of May. We don't try to be late, but we always try like the eighth or the tenth of May, because it seems like you're always itchin' to get a frost around the first of May, so we try to avoid the last frost. If it goes successfully, we're done by the first of June. It usually takes us two or three weeks. I think we plant like six hundred and seventy acres. Two years ago, we got monsoon rains, but we've never planted corn any later than the fourth of July. Planting that late in the season we didn't get enough heat or sunlight in October for it to mature, so the last corn that we planted in July really wasn't top quality, but we chopped it and fed it.

Besides mix-wagon work and feeding the milking cows, John also oversees the planting and harvesting of the corn crop for the farm. The process begins in the late fall, and it isn't simply a matter of grabbing a few bags of seed:

Plate 72: John Clegg adding grain to the mixer wagon.

I pick the varieties in October, November. You order your seed for the following year. Of course we're growing silage corn, so you want to get something that grows a pretty big plant. But also, because of the prices of purchased feed now, grain and so forth, you have to try and get a balance of a good amount of grain in it too, otherwise you end up spending a fortune on purchased feeds. You get most of your energy from the kernels, but you need the fodder too. You have to get sort of a balance there. Because if you get too much of the ear – more of a grain-type variety, then you have high starch, which is good, but you have low digestibility, which is no good for silage. I pick dual-purpose varieties for the most part that exhibit real good silage qualities as far as digestibility and so forth. All fields are different types of soil. Certain varieties will do excellent in one field and lousy in another in the same year – same weather and everything. There really is a difference – you don't just pick out any old corn and plant it. Different farms have better luck with certain varieties than others. I help the people that chop the corn –I drive truck on other farms to help them, or we trade truck hours – they help us, you know, so I get to see stuff there. If I see something I like I'll ask about it and then I'll just try it – a couple of bags here to see what it's going to do. But that's not foolproof either, because every year is different, weather-wise. And a lot of time they won't do good, but you don't know if you don't try. Corn varieties change a little more than you would like. Once you get something you really like you'd kind of like to stick with it, but there's a turnover. They come out with new technology and so forth and want to push something different, and they'll eliminate some of the older varieties. It makes it kind of a pain, actually.

Farmers must always be testing new varieties of corn, knowing that their favorite may disappear in just a few years, but there are also advantages to the new seeds developed, as John points out:

The technology that they're breeding into the newer hybrids – they're trying to get drought-resistant varieties, and then they have the BT corn they came out with a while ago. Cutworms, corn-borers, stuff like that, they can raise havoc with you, but if you plant that BT corn, they'll eat a little bit and then they'll just leave it alone. And then fifteen years ago they went into the Round-Up ready corn which changed things quite a bit too. You could let your weeds get started and then go in and spray while the corn is two

Plate 74: Ledyard Lewis inspects the calf barn with the hired help.

feet tall and it doesn't hurt the corn. Once you spray it with Round-Up, the Round-Up has no residual qualities at all – what is there for grass, weeds, it doesn't matter what kind, it will kill it. But the next day, something new can start. So as long as the corn is big enough, once the leaves shade the ground, for the most part you won't get a real lot of weed pressure after that. The dead weeds also make a little bit of a mat on the ground too, which helps hold moisture in the ground, which some years is good, some years it isn't. This kind of

Plate 73: Annie Luckhardt milking cows.

soil around here it is usually a good thing – it's pretty dry around here. Sandy soils for the most part.

As with most other businesses in America, the larger seed companies are driving out the smaller ones and reducing choice for the consumer. The smaller companies fall behind technologically as well, and end up paying these costs to the larger firms. John says:

I had been buying quite a lot from TA in Pennsylvania, which is a smaller outfit run by one of the Doeblers, I guess, but the price drives that a little bit too. They've had real large price increases on corn seed the last few years, and they're worse than some of the other companies. I think it's because they're a

small company. The big companies, like Dow and Monsanto, I think they make deals among themselves trading technology. There's a technology fee for each bag that somebody else sells too. The little companies have to pay a larger fee because they can't bargain with the other companies. I think that's what happened there. This year we bought two-thirds Pioneer, and most of the other third was Dining Room, which is UAP and CPS merged and they kept the UAP label – Dining Room. CPS used to have Big Row – that was the name of their corn. They just kept the Dining Room name.

To keep on top of what is available and what benefits new varieties might bring to their fields, farmers test them in a clever way:

I'll try a couple of bags of different kinds from year to year if I see something that might work. Because of the fact that there is a turnover on varieties you have to be doing that, otherwise it's almost like starting over again once they eliminate one of your favorite varieties. I prefer to just get two bags, and I'll put that in two of the six hoppers so I'll have four rows of the corn that I've been planting and two rows of the new so you can really compare. In the fields they have the same exact conditions.

The Lewis farm always plants more than they may need for any particular year, knowing that they never want to come up short. As the corn remains viable for years once it is stored away, this can get them through a troublesome time. John discusses how they generally figure out what is needed:

For this farm, we planted six hundred and sixty acres last year. I always used to figure that a dairy farm, to feed all of their young animals and all their replacement stock and to feed the cows for the entire year, you've got to have about an acre and a half per milking cow – that's how I always figured it. That doesn't give you a lot extra, you know, but that will be enough for an average year. That's not including beef and all of that. We plant a lot of corn. We plant a lot more corn per milking cow so we have extra to feed beef, sell it, or whatever. And if you have a bad year, it's nice to have extra planted and you won't sell as much or whatever, but at least you'll have enough for yourself. We've been fortunate – we've always

Plate 75: The large mow used to store thousands of dried hay bales until needed.

Plate 76: Nearly 700 acres of corn are planted to keep the herd in silage throughout the year.

had enough since I've been here.

A typical day for John at the beginning of the planting season differs from his winter routine:

I usually feed the cows in the morning and then start planting. If I have a real nice field, where I can move along pretty good and everything goes well, I can plant fifty acres. We have a six-row corn planter, which works out good because the machine they use to harvest it is a six-row head. They're thirty-inch rows, that's what we plant anyway. It's more efficient when you are chopping to stay in the rows. They have machines that you can go any which way now, but it's still the most efficient to always have six rows going into that thing. We used to put a little bit of starter fertilizer. It drops it in right in front of the

seed. But we gave up that a number of years ago, and it didn't seem to matter. I'd say maybe one out of ten years it would have been a help. Usually if it's wet and cold it would help to have a little bit of starter fertilizer. With the amount of manure we use, we're in pretty good shape with that. I think it was a wise thing, plus it corrodes the corn planter. It saves on that, never mind the cost of the fertilizer.

While it can take quite a while to get all of the fields planted, depending on the weather, it is preferable to have them all chopped at the same time during harvest season, as this is one task the Beriah Lewis Farm hires out. Given the manpower available and the nearly seven hundred acres of corn they plant, it has been more efficient for them to bring in a six-row harvester to quickly get their crops off the field and into their silage pits. Because of this, as John explains, they plant corn with differing maturity rates as they go along:

We usually start planting corn the first week of May. I'll start off planting one hundred and nineteen-day corn, and because it takes longer to get it all planted than it does to go through it with a chopper in the fall, you have to ratchet it down as you're going through the season. It's just a guess as to how things are going to go, you know. But I'll end up planting like a hundred and five-day corn at the end. Ideally, I want the last corn I planted to be ready to chop about two weeks after the first corn planted. That's the goal, because that's about how long it takes to chop.

He had this to say about what he expects for yields with his corn crops:

I think around twenty tons per acre is a pretty good average, but that depends upon how much you let it dry down. You know different people like more moisture in their silage, so that makes a huge difference. You cut it real green and of course it weighs a lot more. A super crop you're up over maybe thirty-five tons today. I always figured real good corn at around thirty. We have trouble with a lack of water a lot of years in the summer – we get shorted on the thunderstorms quite early in the summer here. It's pretty hard for it to rain too much around here. We had an excellent crop last year compared to most farmers. A lot of people were short on corn this year because of too much rain, and it did cause problems in certain places here, but overall it's better to have too much rain than not enough for this farm.

As with all the feed given to the milking herd, it isn't simply a matter of quantity. Each aspect of the food given can affect the quality and amount of milk produced. Corn silage, given the moisture content, can really alter the outcome in the milking parlor, as John warns:

You can get technical about it and take a plant sample and get it analyzed for the moisture content and all of that. You kind of get an idea of what you want. We try to harvest it so it comes out in the twenty-eight to thirty percent dry-matter range. That works the best for us. If you get it too dry it's hard to get them to milk as well. A lot of people will let it dry down because the grain dries down more. You get a little bit more starch in the corn, but you lose so much on the digestibility. It might not even show up on paper, but if it's too dry, I don't know if it just doesn't go through their system fast enough, or they can't eat enough of it, or what it is, but it's definitely the case for here. Better to be a little too wet than too dry.

Plate 78: Corby Steadman moves a team of oxen out for exercise.

Plate 77: Number 550 gets her feet wet.

Corn silage makes up the lion's share of the feed ration – some eighty pounds each day per milking cow – but the grasses raised also form an important part of the mix. They do not, of course, need planting every year like the corn, but instead may be replanted every ten years or so as long as they keep spreading manure on them. John discusses the composition of the grass fields:

I guess your regular old hay fields they seed with quite a bit of orchard grass. The hay that we want for the milking cows we prefer to have that mostly alfalfa. It will usually be an eighty/twenty, alfalfa/orchard grass mix, because it's harder to dry one hundred percent alfalfa. You need four good days to do it, or three super days.

As Ledyard describes it, harvesting the grasses they need is another time-consuming task for all of the farmhands:

Grasses we always try the middle of May to get a four-day window – we have for the past five or six years – where we can dry hay in the middle of May. That is real early. A lot of guys start around the first of June. My father used to say any hay made in June is good hay. So we go through middle of May through June to get first cutting off. If we get adequate rainfall throughout the summer, we try to go every thirty to forty days and get another cutting. So we have gotten five cuttings off in one year, but three cuttings is normal. Sometimes you get four if you can stay on top of it; if you're real aggressive and put fertilizer on it after the first cutting and keep going you can get five. There's orchard grass, we have alfalfa, we have just regular grass.

Not only do they use different types of grasses, but they are also separated by the time of year they are cut for quality reasons before they are fed to the cattle. Ledyard goes on to explain:

There's a lot of different qualities of hay. The first cutting is usually the most quantity. Second, third, is really the best quality, and then when you get to the fourth or fifth it's getting later on and into October and that's pretty nice stuff, but it's not like the second – it's one step down. First cutting is always the most quantity. If you fertilize the field, the saying is you get a hundred square bales per acre, first cutting. That's very possible, so if you've got a fifty-acre field, that's five thousand bales.

Plate 80: Ready for another bale.

Additionally, farmers may grow cover crops over newly chopped cornfields. According to John:

We mostly use rye for cover crops. One year rye was real expensive, so we planted winter wheat, but last year rye was cheaper than wheat so we went back to the rye. They mainly stop erosion because of all of the hills around here. You do get organic matter, and it's a good thing that way too, but mainly it's the erosion.

As with any crop planted, farmers must also worry about weather and pests dimming their futures. While as of yet nothing can be done about the weather, pest control has also been engineered into the corn itself through the use of genetics. John had problems with black cutworms one year and used special seeds with good result:

I did buy Dow AgraScience, which is Hurculex technology. And it works – I tried it and they didn't even touch it. You could see that there was pressure from the cutworms there, but the corn was all there. The year before, we lost I'll bet you, I don't know, in the one field, I'll bet you thirty, thirty to forty percent. They put a hurtin' on. They get it when it's small. Before I knew a lot about it

Plate 79: Elizabeth Lewis brings another load of hay to be stored in the mow.

I could see there was a problem, but I didn't know exactly what it was. We had a couple of guys come down and walk the fields, and of course that's what they do, so they knew what the problem was. You can spray for them, but that costs a lot of money too. And of course that's an after-the-fact thing, By the time you call somebody and get the sprayer here and so forth, they can do some damage. I don't know, must be millions of them. They cut it off right at the base of the plant. They don't necessarily cut right through it, but they eat enough so that it kills the plant. It slowly withers up, then it falls down and blows away. It will look like the planter didn't plant – like you had planter troubles. All kinds of skips. But I look at it enough times that I knew it was there – it came up.

Plate 81: Chopping corn with a six-row chopper and a fleet of dump trucks.

Putting all of your eggs in one basket, or in this case, all of your milk in one pail, has its dangers, of course. Ledyard and Ted have been able to diversify somewhat given the time constraints they face by hiring extra help on occasion. Ted lists some other ways they help make ends meet:

We sell our hay, we sell silage, we sell wood, we sell beef too. It's tough. I don't know how people do it on just a milk check and a beef check, and ninety percent of the farms do. We do a lot, but there's a lot going in and a lot going out. I would say our milk is still ninety percent. But to have the help we have, you have to do other things to justify it, and to keeps things up – like painting. It's quite a bit just to get the walls painted. In the summertime we have kids haying and painting.

Beriah Lewis Farm also has a significant beef herd to go with their dairy animals.

We have beef cattle too, and we have probably a hundred and fifty of them, so you're looking at eight hundred and fifty head of dairy cattle. If you drive by our beef lot, they look nice, they look good, you know. There are some fat ones over there, but you've got to have the corn silage, you've got to have the hay to put into 'em – mainly the corn silage. Out West they grow a lot of corn and they just pump it into them, and then you have the beef. My brother and I, besides milking cows, we've got a lot of small sidelines. We branch off. Some of them are pretty good. It's just like the economy – sometimes it's pretty good and sometimes it's pretty bad. The beef market has dropped quite a little here, in the past six months to a year, but now it seems to be rebounding and I hope we can take advantage of the rebounding here pretty quick. The USDA has a thing, that if they go through a USDA slaughterhouse, to be prime they have to be under thirty months in age. We buy them when they weigh four to seven hundred pounds, I would say, and I usually keep them until they weigh fourteen hundred pounds. On a good one, that usually takes about a year. I usually buy them at a year old, and I keep them here a year and a half and then they go to the slaughterhouse. I sell a lot of 'em privately, but then I do get an abundancy and then I take them to an auction.

Both brothers have other pursuits:

My brother has a sideline of oxen, and I have a sideline of trucking, and I take a lot of animals to New Holland, Pennsylvania, and then I take some to a little auction in Fairhaven, Massachusetts. The auction in New Holland Pennsylvania is huge – beef, dairy, horses, goats. I see last week they ran nine thousand head through it. I back up with a tractor trailer into the one in New Holland and they unload 'em and it don't even put a dent in the place. My brother went down a couple of months ago and he said that the tractor trailers were backing in there full and tractor trailers were taking them out of there. Unbelievable how many cattle they were handling. A year ago when milk was high, a good animal

Plate 82: Curt Main packs a mountain of corn silage in the fall of the year.

was bringing – I know a guy that sold a trailer-load of cows for twenty-two hundred dollars apiece. Now you're lucky to get eleven hundred for an animal – it was cut in half. Same amount of feed, same amount of time.

John isn't certain what the future might hold for farming in Connecticut, although he feels confident about the future for the Beriah Lewis Farm:

I don't know. A farm like this has been here forever, practically – it's well established. It's pretty impossible to start a new farm right now in this town I think. I think there is a future for a couple of them anyway. I never really look at it like I have to go find something different because I won't have a job here – I don't feel that way. I do feel there will be a job here as long as I want to do it and they want me to do it.

And for him, farming is in his blood, and the complexities he faces every day help keep the job interesting for him:

It's not just the dumb old farmer down the road anymore. There's definitely more to it than most people realize. That's also what makes it attractive, because you're not doing the exact same thing, day after day, hour after hour. In a day I'll go deliver a load of feed, I'll feed the cows, I'll go work on the fields, you know? You're always doing something different. When it's nice out, it is awful nice to be working outside. I never had an inside job. Just from being in school and looking out the window I can still remember that feeling. I didn't like being in that building, sitting there.

Like Ted and Ledyard, Rosalind has an obvious sense of pride in all that the family has accomplished in the continuing operation of the farm, as do other generations of Lewises living elsewhere in the country:

The Lewises were very proud of their heritage. That was one of the reasons that I think the farm has lasted this long. There is one left of the sixth generation – she is ninety-six years old, and every birthday, and once during the summer, she gets somebody to bring her to the farm. And then there are my children, and one left of the next generation, and she's eighty-four or five, and she's in Seattle Washington, and she comes every year and checks what's going on at the Beriah Lewis Farm. The boys have inherited that pride of their heritage, and the girls too, but of course the boys have the name.

She also notes that things have changed not only on the farm, but in the country as well. The work ethic that has always been central to each of the farming families in town seems out of place in today's world:

Now you have more help, and help is not as dedicated to what they're doing as you are, because they don't own it – you do. We didn't have a lot of help in those days, and then we did have help,

Plate 83: The double-eleven milking parlor.

but it was the family and help. Well now we have help, and the family can't be there at every step of the way, and I see the difference there. We had workmen's comp and unemployment, Dad and I. You had to pay it – it was always a tax – but I never filled out a workmen's comp I bet until the last ten years. And now they stand right there at the door and say, "Do you give a pink slip?" So it's the difference in what you want out of life. There's work here. If you want to do it, there's work here. I have one that isn't family, full time. We have a lot of part time that come and go. Again, you cannot have them on the payroll if they're not dedicated. It costs you money. See Dad and I didn't have to deal with those things. There were fewer animals and fewer people. When the children were in school, we had two full-time men that worked with Dad. They were almost as dedicated as Dad was. It's a whole different science, working. It used to be that if they finished chores, they would paint the side of that building or they would cut brush. You're not going to have people do that anymore. Or they put the stones on the walls in the spring when the heave was over. That isn't there now anymore.

Ledyard reflects on the same phenomenon:

It seems to be a different breed of people who want to be on a farm. If you get the person that's gone through four years of college and can't find another job, and you try to get them on the farm, they just don't seem to have the ambition. Our help turn-over ratio isn't too bad – John has been here thirteen years, another one's been here five years. Butch has been here twelve years. One fellow left us and he was going to go make twenty-five dollars an hour selling vacuum cleaners. He left and we hired somebody else – now he wants to come back. "How many vacuum cleaners did you sell in three or four months?" He said, "One." I'm seventh generation, and I don't know

Plate 84: Craig Graves loads hay for the loft onto the conveyor.

what the work ethic will be with the eighth generation. Right now I have two nieces and three nephews that could work for us and be with us, and there's only three here, one on summer break from college.

Beyond the issues of hiring help as the farm expanded, the costs of doing business have also changed as the farm grew through the years. Ted lists just a few:

Help and insurance and workman's comp. Grain's the biggest cost. Grain is a third of the milk check – that's a lot. Between grain and the payroll, it's quite a bit. Then you've got insurance and taxes. You still have the payment on the tractor or payment on something. You don't buy unless you can pay for it. In September, in two or three days you're going through a thousand gallons of fuel when they're chopping corn. That's why we sprayed. You look at how many gallons it takes to do an acre, on top of your grain bill, on top of your payroll, on top of your insurance, electricity – you've got to be pretty tight to get by. The bigger you are the bigger the bills, but the smaller you are the milk check is still smaller, so it's pretty tough.

Rosalind explains the harsh economic realities facing North Stonington's dairy farmers:

It's a very, very difficult business. When Dad died in 1997, milk was about nineteen, almost twenty dollars a hundred pounds. The following September, it went down to twelve dollars a hundred. And last spring, we were making ten dollars a hundred, and Dad's been gone thirteen years. I think our last check was about fourteen. But that's not enough. You can't break even. When Dad died, we were doing pretty good. Dad said you could make money at seventeen- or eighteen-dollar milk, thirteen years ago. Well you can imagine three families, the boys, me, and help at ten dollars – ten-dollar milk.

Ted also wonders about the future for the once common, smaller, family-owned farm. He feels that where his farm is now may be the size at which it should remain, at least until changes occur in the dairy world:

The small farmers, to milk under a hundred cows is tough. Over the last thirty years, people who milked a hundred, most of them need to be milking three hundred or four hundred now. We used to milk a hundred and fifty, a hundred and eighty, but now we're milking three thirty. Land's a problem too. With everything else we do, we couldn't milk five hundred and depend on milk more. But with the wave of the milk, the ups and downs, that doesn't seem like the way to go. The future just doesn't seem like it was thirty years ago, to invest that much into milking another two hundred cows. That's my opinion. My brother-in-law – they milk a thousand cows. They're buying cows all the time right now. I can't see that, but all farmers look at things different. Whatever works, works, you know? I'd rather do a good job with three hundred than milk four hundred. There are some guys that milk four hundred and make the same amount of milk that we make milking three hundred. You can look at it that way too. These people that milk a thousand cows could milk seven hundred and make the same amount of milk. Then there are people that milk five thousand – it's crazy. Before we built the free-stall barn we had two hundred. Overcrowded. We were milking cows twelve hours a day – twice a day, six hours a shift – those two hundred cows in a double-six parlor. There was a hundred and ten free-stalls, but we were gearing up to switch into the new barn. That's probably the most money we made milking cows. We didn't have extra help, you know, we probably only had four guys and one barn. Not a lot of overhead. To build the new place twenty years ago – that was quite an investment – and now to run it, you have more employees, more electricity.

Plate 86: Young calf in the calf barn.

Plate 85: Catherine Lewis – eighth generation.

The largest part of the problem with milk pricing, as local farmers see it, comes from the West, where milk is produced on a huge scale and shipped to the East to compete with milk from the family-owned farm. Ted talks of the scale unheard of in New England:

I've been to California – you can own fifty acres and milk a thousand cows out there. In California, they don't need barns – it doesn't cost as much to make a hundred pounds of milk out there, but still they flood the market. They say in Texas guys are milking twenty-five thousand cows just because they can buy all their feed. There is so much cropland they buy all their feed and make their milk. The little guy is going to get pushed out. It depends on if you want to be little and work every day and be dedicated and that's what your purpose is in life and you love doing that, but three hundred cows is small now, where it used to be big. Twenty years ago three hundred cows was a big herd. Now three hundred is probably the borderline of making it.

The farmers in town, when asked if they see a solution to the problem of low and fluctuating milk prices seem to feel a quota system may be the answer. Knowing that they can produce a certain amount of milk and be paid a certain price for it would take the random behavior out of the marketplace and give farmers a sense of what they should or shouldn't do to keep their business viable. Ted feels that:

There needs to be a quota. Canada has a quota, and they don't seem to have the variation. They know they're going to get seventeen dollars for their milk right through the year. Where, like I said, these guys are going to throw another five thousand cows on if milk goes to twenty-four dollars. Quota is per cow. You're talking maybe five thousand per cow. So you're going to put on another hundred cows? Right now they're picking up twenty-five thousand pounds of milk here. Say the United States government is going to have a quota. Well, if we ship twenty-six thousand tomorrow, we're going to get paid eighteen dollars for twenty-five thousand, and we're only going to get paid twelve dollars for the thousand pounds of milk over. But say Niles is going to sell his cows, I can buy his quota for sixty cows, and that's how they work in Canada. And I know it stabilizes it more –

Plate 88: Young hands exercise the oxen every morning.

you don't have the high and lows – the fourteen-dollar milk and the twenty-four-dollar milk. Somebody that milks sixty cows, you know they've got that quota and they're valuable. It's better than money in the bank. If they want to sell, sell, but they still know if they're going to buy a new tractor they're going to get eighteen or twenty dollars – whatever the quota is. But if you go over your quota you're not breaking even. There is a penalty. What I understand about it, a quota would work.

When milk was high, everything else was real high. Then milk went down and everything else didn't come down. I don't really see how in the future – fifty years from now – I don't see how you're going to get a person that dedicated, like the Miners. There is that small window of profit. They're established like we are,

Plate 87: Timbers in the loft in the old section of the barn.

but if I wasn't the seventh generation, I don't see how I could be in business today. And a lot of farms went out of business. We always thought land was going to be a problem, and it will be fifty, a hundred years from now. They don't make any more of it and it's going to get developed. The old Brown Farm, Ackleys, Dennison Farm – all those farms we rent – they used to milk a hundred cows and were successful, but the next generation came, "Do I want to do this seven days a week?" It's not that they couldn't do it ... .

As many of the farm owners in town have stated, having had the farm in the family for generations has been a big advantage, and few see how new farms will ever be able to make a go of it if they have to acquire the land on top of everything else needed to run the business. This has not, however, kept new and large-scale farms from springing to life in other areas of the country on borrowed money. Ledyard talks about these farms living beyond their means in an effort to survive:

We had the banker come and tell us there's people that are $5000 in debt per cow. I went to a farm in Wisconsin a couple

Rosalind wonders about the future of Beriah Lewis Farm, knowing that the next generation's aspirations lie somewhere beyond the farm's stone walls and fences.

If they are going to stay in – not me – I'm pushin' eighty, in my lifetime I'm done – they have talked about things that they'd like to do. They'd like to diversify and still keep the farm. I really don't know how they will be able to continue, the next generation, making milk. Now I could be all wrong, but you know it's twenty-four/seven, and it's three hundred and sixty-five days a year. And I can see it in the grandchildren. The dedication is not there. Oh, they'll get it done, but not the way we do it. Some of the grandchildren are really interested, but you know, interest is one thing. You have to have a living. When others are making twenty dollars an hour and you're only making two, if you're the owner, or maybe some weeks you're not making anything, then you have to think twice. It's been a wonderful life for my family. We had fun – not a lot of money – but we had a lot of fun. They love it, as you can see, but there are days around here that are pretty tough. Pretty tough.

The work on each farm in town is unending, and now, more than ever, the future is uncertain, but the dairymen and women forge ahead doing what it is they love. As Ledyard summarizes:

We milk from 5:30 – 10:30, 1:30 – 6:30, and then 9:30 at night to 1:30 in the morning. There's twelve of us on the payroll. Some of us work seventy to eighty hours a week, and some of us work fifteen. A guy followed me around for a day and he said, "You know, you've got a hard life, but you've got a good life," and I said, "Truer words were never spoken."

Plate 89: Ted raises and trains oxen for pulling competitions and is one of the best in his field.

of years ago. They're milking 1400 cows. They're twenty-four-hundred dollars a cow in debt. They had plans to make their own electricity from methane. They were taking all of the liquid out of the manure – separating the water from the manure – it was pretty neat. I went into another room and there was this generator there that was this high, and it would go from here to that washer and dryer, and they had blown it up, and the heads were on the floor, and they said it was like two hundred-thousand dollars to get new heads for it. The logistics there – it never looked like they were ever going to get ahead. There's a lot of farms like that.

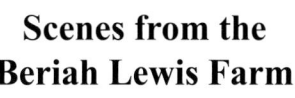

**Scenes from the
Beriah Lewis Farm**

George & Carrol Miner Farm

Plate 90: View of the Miner farm and barn complex from the adjacent hayfield.

The George and Carrol Miner Farm rests atop Chester Main Road, one of the higher spots in North Stonington. Before surrounding trees grew too tall, views of the ocean some ten miles distant could be had on clear days from the hill. The farm is owned and operated by Carrol and Betty Miner, their son Orrin, and their nephew Robert Miner – the son of Carrol's brother George and his wife Frances. The homestead farm was down the hill at what is now Cool Breeze Farm, as Carrol and Niles are cousins. Their fathers were raised on Cool Breeze Farm, and Carrol's father eventually bought the farm they currently occupy. Carrol himself was born in the current farmhouse in 1928. He says:

I had two brothers, George and another brother, Maurice – when he got married he moved over on Wintechog Hill. He farmed it up there – he was the oldest. My father's homestead is down there where Niles is. He bought this farm in 1919. He bought it from Will Babcock – he lived down there where McGowan used to live, down there on Babcock Road. He owned that farm and he owned this farm.

He said he could never make a living here, so father bought it, then he moved up here. He did dairy, and I guess a little bit of everything. He did a little wood business, 'cause he used to tell about haulin' wood to Westerly on the old cane wagon. I don't know about beef cattle, but I know they used to raise up stuff so they could feed themselves with it. They used to keep a few hogs, I don't know how many. I don't know if they ever sold any of 'em or not – I never heard him say.

Before mechanization, farmers in town depended upon animals not only for transportation, but to help with the fieldwork as well. Carrol recalls working with the farm's horses and the Cletrac, the first machine they bought:

I think they had two teams of horses – I ain't sure. Just plain work horses, that's all. I never heard him tell about usin' oxen. Oxen are slower, more steadier than horses. All we ever used was horses. I drove horses, raked hay with horses. I did mow a little with horses, moved some hay with horses, where it wasn't too stony, where it was more smoother. We bought the old crawler there in '39 – bought the old Cletrac. Then they bought the 3N there, the International, the one I saw wood with. I guess they bought that around '42. Some guy had it, he had to go into the service, and he didn't know whether he was goin' to come back or not, so we bought it from him. He came back, but he never farmed it after he came back – well he did farm it for just for a short time, that was all.

In his father's day, the dairy operation was smaller than it would eventually become. Many factors in the equation at that time limited what they could either produce or sell – lack of modern refrigeration and transportation being just a few. Carrol remembers:

We had maybe about twenty-five, thirty cows. Milkin' around fifty right now. Was up to around a hundred, but we're milkin' fifty now. You'd sell milk in cans – it went to Providence. Albro used to pick it up. Before that, they used to take it down to the village and put it on the old trolley and it went to Providence that way, 'cause the trolley went through Westerly, and they take it to Providence. You used to cool it with ice – put it in water and put the ice in it and cool it that way. They used to say a good cow, back then, if she give sixteen quarts, she was a heck of a good cow. Today they got 'em up so they milk, some of them milk a hundred pounds or better. Not all of 'em, but some of 'em. They bred 'em up more, and they feed 'em more grain now. Back then, the grain you got, all your protein, everything, wasn't as high as it is today. Used to buy it all in bags. Dairy feed, the cows used to eat. Father used to mix his own grain some of the time there. He'd buy some bran, and some chops there – that'd be ground-up corn where they grind the ears, cob and all. Some time he'd feed just a little grain with it – mix it, or he'd get some gluten – put it in it. Sometime you'd get it from Campbell Grain Company down in Pawcatuck, sometime you'd get it from Big Y, and sometimes he'd get it from Eastern States. They went into Agway afterwards, but it used to be Eastern States. Wherever they could buy it the cheapest. He'd have to go get it, then they got so they would deliver it in bags. Never heard of bulk back then. Bulk started about the middle fifties, I would say. We're buying it from Ventura right now. One time we were buying it from Central Connecticut, 'till we run into trouble with 'em; then we switched grain.

The planting of crops in his youth was much different from the way in which crops are planted today throughout town. As with all work, farmers came up with ways in which to

Plate 91: Double-four milking parlor.

Plate 92: Carrol Miner.

save labor in the planting and harrowing of fields, as Carrol describes:

We used to plant a little millet in the midsummer and cut that off – pull it in with the old horses and team wagon – feed it to the cows, and just mostly timothy hay, and red-top, and clover. When they seed it they'd mix it together. To start off with, they'd seed some oats down, and put the other seed with it. The oats would come first. You'd mow them off, and then the next year you'd have the other seed there comin'. You didn't have to harrow the ground back up again. Sow all the seed together. Sow the oats first, 'cause that was a bigger seed, then you'd sow the other

Plate 93: International tractor by the shed.

seed at the same time. The oats, we used to harrow them with the old horses, light, then the other seeds you just used to bush them in. You had a plank, it would be probably eight-foot long, and you'd drill holes through it, and you get white birches and put them through the holes, hew 'em down enough then get 'em through the holes, you put a spike down through 'em, that would help hold 'em from workin' back out, and you'd drag that over the ground – just enough to work the hayseed there, 'cause you didn't want it in too deep. A couple of inches.

If you didn't have manure, you'd buy fertilizer and put that on. We used to have an old fertilizer sower. You'd put it on, it had a couple of bins on it, you'd pull that with horses. Kind of hook onto the back of a wagon, and you could have the fertilizer on the wagon. One drive the horses, the other one keep dumpin' the fertilizer in – shovelin' it in while you emptied the bag. All used to come in burlap bags.

As his wife Betty notes, while they worked to save labor, they also worked to keep waste to a minimum with everything else on the farm:

Some people used to take the burlap bags and make hooked rugs, through the holes of the burlap. The cloth bags the grain come in, they used to make pillow cases, and make things out of those.

But Carrol reflects on how those times are long gone:

Most of 'em all plastic now – it's work to find a burlap bag today. That plastic stuff – they won't take 'em back. They say they can buy 'em cheaper than they can clean 'em up, fumigate 'em, 'cause everything's got be fumigated so they don't take one disease from one farm to another. Just throw 'em away. Ain't nothin' wrong with some of 'em.

Carrol began with chores, as did all children on the farm at that time, early in life. He also attended the one-room schoolhouse at the end of his driveway. While it was easy to get to, as a child it had its disadvantages:

Well we didn't have much fun, mostly work. When we was goin' to school, we used to go out to recess and play tag, or hide-and-go-seek, or kick the can, or somethin' like that. We had chores to do, just the same as everybody else over there. We didn't have much time to play. I did chores ever since I've been big enough to do anything. Used to have to go up in the silo and throw the silage down, sweep up in front of the cows, sweep the feed into the cows. Yeah, I've done it ever since I was big enough – as they say – since I could walk. So that's all I ever knew what to do. I went to school one year down here, right across the road. You'd have to behave yourself though – your father was home! Many's the

Plate 94: Road to the back fields on a foggy summer morning.

time, go out for recess, he'd be out there pickin' stone or somethin', he'd say, "Come on – come on out of there and help pick stone!" Went out here one year, then they closed it up, 'cause there wasn't enough of us to keep it open. Carrol Maine went to school over here. Coats, the old folks, they all went to school here. I only went here when they had a woman teacher, but there used to be a man teacher here. One room, all eight grades. Then I went to Center School, down there by the town garage. Then they had the Wheeler School downtown that they built. My oldest brother graduated from there.

There was little time for children to get into mischief around his father's farm. Carrol continues:

Sometime you used to go to school pretty tired. Of course you didn't have homework from school then, so when you get home you had to change your clothes and go out and help do whatever they was doin'. If they was plantin' in the springtime, or summertime you'd be hayin' it, or if you get the hayin' done, you get done before somebody else, you go help them out. Same way in the fall of the year. Cut corn by hand. You'd have to cut it by hand, pick it up by hand, throw it on the wagon, and run it though the old silage cutter into the silo. If you got done before somebody else, you'd go help them out, or if they got done before you did, they'd come and help ourselves. One hand would wash the other.

As Carrol grew older, new

technology was bought to make the jobs they did easier. Each farm couldn't afford, however, to buy all of the new labor-saving devices at once, and often they bought different pieces of machinery and shared them among themselves as they did with labor. Carrol remembers how buying new equipment often led to more work for his father:

He would do a lot of outside work. Harrow for different ones, or pull stone out of the fields for different ones. Help keep things goin'. You had to work them days, you didn't sit around watchin' television – didn't have television to look at! Father used to work right up as long as he could see up the lot.

Helping the surrounding farms and receiving help from them in turn allowed each farm to prosper and kept a sense of shared purpose and common goals within the community. Carrol recounts how corn was harvested long ago:

I remember cuttin' corn by hand. I was only seven or eight years old. A corn knife, somethin' like a sickle, only a corn knife has a handle on it about so long and the blade kind of a half-moon. Cut the corn and lay it in bundles. The whole stalk and everything was all chopped up with an old silage cutter – went into the silo. It was run off an engine, or run off a tractor off of pulleys. Anything would drive it. Some of 'em used to use them old two lungers on it. The worst part of them, the belt would flop up and

Plate 95: Carrol and his nephew, Robert Miner, milking cows in the early evening.

Plate 96: The older stanchion barn and shed complex. The milk storage room is to the left of the car.

little extra work for the women on the farm in the form of meals for the itinerant crews:

They used to help out Uncle Frank and, like I said, they used to go down and help Bankers. They used to go around. They was real sociable – go down the hill and do a lot of work. When Carrol was little he used to carry water down to 'em, so he told me. Or carry 'em down their lunch or somethin' else. But usually the people back then used to make meals. The women folks then. If the men was workin' around there and stuff, they'd have a big meal for 'em. It wasn't just a soup-and-sandwich deal like some people do now.

The new technology farmers began to experiment with in town did take some getting used to. Carrol Maine, who owned the farm next door to the Miners, bought one of the early tractors in town, and as Carrol mentions, was anxious to show off its capabilities:

That was the first tractor he ever owned. They had tractors here in town. He was up here in back plowin', and he wanted Father and I to come up and see him plow. So we went out and watched him. And he come across the lot, and he come up to the wall there and he hollered, "Whoa!" Tractor didn't stop – went up the wall – the old plow stuck right out into the ground – he couldn't back up and he couldn't go ahead 'cause he was up against the wall. He finally wiggled around and he got out.

down just like the John Deere tractor with the two cylinders. My Uncle Palmer down there, he had one. My father and his father, they used to work together. That'd be Niles' father. They'd go down there and do the hayin', then they'd do my Uncle Herbert's hayin' – he lived down there where Lattimore lived – then they'd come up and do my father's. Then, years later, my father got his own, and did his own cuttin' corn and hayin'. When we got the first corn chopper there, power take-off, we used to cut corn, go down cut Frank White's, down where Richard is, fill his silo, then go down, cut Bromley's corn, then go over and cut my Uncle Frank's, cut my brother's over there, then if we got done, sometimes we'd go over and help Lewis' chop their corn – Harry Lewis. We used to chop a lot of corn.

Betty Miner recalls that when the men went to help neighboring farms, they inadvertently created a

Before the advent of the modern pit silo, where three large concrete walls form a box for storing chopped hay or corn under plastic, standing silos were used to keep the feed for cattle until needed. Carrol also talks about the other feed materials they had to purchase, not being able to produce them on the farm itself:

Father would plant some corn – enough to fill the silo. We had the old upright silos, wood silos. The silo company would put 'em up. He had a Unidello silo, and he had a

Green Mountain silo out there. The Unidello was the best. The wood was cypress or fir. Some had metal roofs, some had wood. Ventilator up on the very top on the Unidello silo. Corn, sometimes there, if he run short of corn during the springtime, they'd chop up some hay and put that up. He used to put some molasses with it. It would help it ferment and also give it a little more nutrients. You'd get it from the grain place. It was in barrels, fifty-five gallon drums. Now you can't even buy molasses, not real molasses. You can buy a blend from concentrate, 'cause we give that to the heifers out here – it's got minerals and stuff in it. But you can't buy straight black-strap molasses. We used to get citrus pulp to feed the cows. It used to come in down there by railroad car. And beet pulp. And you had to get that unloaded by a certain time. That was in one hundred-pound bags. Them old beet pulp bags was big!

Even life as a citizen in town was much different. While today we wouldn't think of going out to help the town road crew, it was just part of life in North Stonington, as Carrol remembers:

That old town buildin' down there, my father helped build it. Pa said a lot of 'em would get together and help the town out. He used to go out here when they used to mow the roads, if he got through hayin' in time in the summertime, he'd go out and help 'em mow the road out here. And if we had some stone to blast or somethin', they would come up and blast 'em. They wouldn't charge anything for doin' it because we'd trade work for the town.

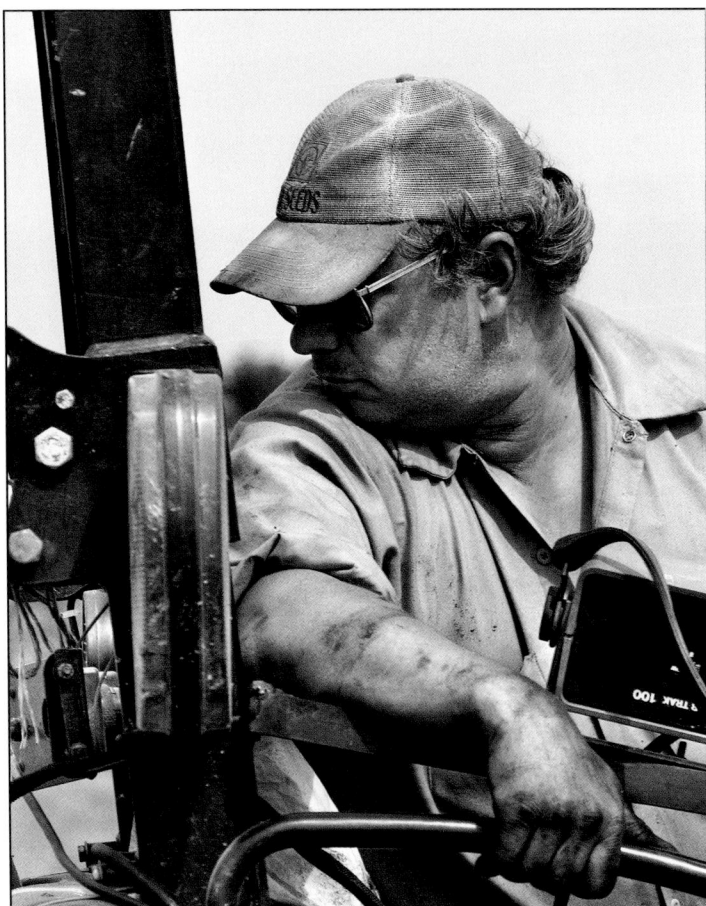

Plate 98: Robert Miner baling hay in the early summer.

Even the weather seemed harder in New England years ago. Although it is not uncommon to see deep drifts of snow up on the hill after a major snow storm, it doesn't compare to one winter Carrol remembers very clearly where the snow was piled next to the power lines above:

Had a picture of the snow up the road here, that year when we had so much snow, when Woody Klewin was up there pushin' it out with a dozer. I have a picture of him sittin' on the dozer there, reachin' up touchin' the wire. I would say somewhere around '42. That's the most snow I ever seen. Even the driveway out here was filled in, from one wall to the other, right level with the wall. Had one snowstorm right on top of another. Father had to take the old horses and go out through the lot there with the sleds, take the milk,

Plate 97: Carrol baling a field of hay.

over to Carrol, picked up Carrol's milk, and they carried it down to the village to get on the truck that year.

Betty and Carrol knew each other for many years before they married in 1957 and began their family of five children. While Betty didn't come from a farming background, it didn't mean she was completely unfamiliar with some aspects of farming life:

I was born in Richmond, Rhode Island. Well, it was Arcadia. In one of those mill houses. My mother worked at housework, and when they

Plate 99: The farm's bull rests in the cool barn on a hot summer afternoon.

was in Westerly, they worked in the mill. But then we come up country. That's what I used to have to do when I first started in housework, and babysittin' and certain things. They had a cow, just for milk and stuff, but they weren't in farmin'. I knew Carrol from the time, you might as well say, that I moved up country. I knew him for about thirteen years before we got married. And how come I knew him was through the 4-H and stuff from my uncles. They used to go around different houses, like the boys come here when the mother and them was livin' – have the 4-H meetin's and come up to the house.

Before she and Carrol were married, Betty was working in one of the largest textile mills in the area:

I worked up to Baltic when we got married, in a weave shop. They didn't want me to quit from up there, but he said no, it was too far, and he says he was marryin' me, and if he couldn't support his wife, he wasn't goin' to get married. He says, "A man's supposed to support his wife, not make his wife work." It was a big mill. I was fillin' bobbins. I would put 'em in the thing in that and they

was already wound up – those was. I have wound bobbins when I worked in the Fall's Mill. That's where I was workin' when I got married, and they didn't want me to quit because between, oh what was her name – I can't think of it – between her and I they was gonna send us down south for a race contest, for fillin' bobbins. Well there was a thousand machines in that shop in Baltic. Well of course some of 'em wasn't runnin', but the biggest majority was, and you'd hear that clang, clang all day long. You'd come out sometimes and you almost hear it still clanging! I'll never forget the time that they had the fire up there. I was from here to about there – to the door – and all of a sudden, clap, clap, clap – just like that, three machines caught afire. And it was hot. It was a hundred and thirty-five in the weave shop. And by the time it got to the third one, the men had it out, because there was firemen that worked there and they had these buckets that was on the poles. They had it out. They was quick, but boy it scared me so all I could do was stand there. I couldn't even move – I was paralyzed! Where the machine works, you know, it makes that lint, and of course it 'cumulated, and I suppose the machine got hot, and hot just enough that it made it catch fire. It was dangerous up there though in some ways.

Betty describes her early childhood years as quite difficult, and when she told her grandmother she was going to marry, her grandmother had concerns:

So she says, "You know what you're doin'?" And I sez "Yes." And she sez, "Well you know, farm life is hard." I almost said, "Well it ain't no harder than what I got right home here." 'Cause I used to have to do a day's work before I went to work, work all day, come home and work more. I painted at night when I come home, from work and all that. Painting, papered. I was twenty-eight when I got married. Moved

Plate 100: Orrin Miner

up there, and I had to be here to cook and stuff too, and then I helped him in the garden and used to go fishing with the mother and do different things, you know? Help take care of the house and everything. I took care of his mother and father. I'm not sorry I did it though.

Raising five children and helping with a variety of chores around the farm left Betty with little or no free time. While the farm was a great place to raise children, it was also filled with many attractive dangers:

Get the children up, get 'em ready for school, but when they was real tiny, then I had to be with them a lot, and then I used to watch 'em on account of the machinery and stuff. Like they had the auger and blew corn up into the silo out back when Marilyn was real small, and it used to scare me 'cause I was so afraid that they'd get hurt. They're curious. Helen got kicked out of the barn door by a cow one time when she was little. You should have seen her head, and two days afterwards she had the most beautifullest shiner you ever saw. I used to take 'em up fishin' and up back and down here and catch bullheads. He used to love his bullheads – soak 'em, and then go ahead in that and cook 'em for him. I used to go out fly kites with 'em, slide down hill, all that stuff that a mother does. Because he'd be busy workin' or somethin'. He didn't get much chance. He left it all up to me, same way as school and stuff. Anything goin' on I had to do it.

As the children grew old enough, they would help Betty with chores outside of the house as well:

The girls used to help. The cows was down across the road, and the girls and I used to drive them up when they was out in the fields doing their work and

Plate 101: A hayrake sits in the newly mown field after use.

that. We used to drive 'em up, put them in the barn, and even start milkin' sometimes. I know at one time there he was milkin' over a hundred. Same was as before George died, he was milkin' right up there in that – I think it was a hundred. The milkin' parlor and the free-stall barn come about the same time. The cows used to come up the driveway, and the girls and I used to go down and get 'em and bring 'em up and start the milking lots of times. We had a dog then. One of the other dogs was pretty good at drivin' the cows.

As with many farm wives in town at that time, a large portion of the day was spent in the kitchen. Betty learned to cook, in part, not only by helping her grandmother bake, but also by helping a neighbor when she was still a child:

I used to help make pies, make different pies. I always helped my grandmother bake – she's the one that helped me. And then Mrs. Sheffield. When I was only, what? Seven, I guess it was. I used to go over – she lived right next door – and she taught me how to cook and that because she couldn't get around and everything. She was sick – had dropsy, so I used to go over there and that and she would tell me what to do and stuff like that, so I learned how to cook there too, because Bill and Frank used to come home from work, and I'd have supper all ready for 'em. I've always had to work. I remember standing on the counter to put the dishes up in the cupboard for my grandmother. Barefooted – stand on the counter and put the dishes in the cupboard, and I'm still workin'... .

To accomplish all that needed doing each day, Betty would have to rise early to begin work. With all of the manual work to be done, appetites on the farm were larger than they are today, and fueling the workers was a full-time job in itself.

Betty describes some typical meals from the early years:

Get up in the mornin', and if you was goin' to have a boiled dinner, either you used a hock, or you use some ham, but the hock's just like ham, and you put it on, got it to boilin', then you'd put cabbage, carrots, potatoes, turnip if you had

Plate 102: Betty Miner working in her vegetable garden.

it, and make it a real boiled dinner. Because at noontime, when the other ones was workin' with ya, and workin' around, they would come in and eat. So I always had pies, or cookies or stuff like that for dessert. But in the mornin' was usually, like I said, either boiled dinner or it would be, well, lots of times it was bacon and eggs, and you know,

Plate 103: The old milk house.

the regular breakfast. Or sausage or somethin' like that. Johnnycake! I can't forget johnnycake – three meals a day when I first come here. Mornin', noon and night – johnnycake. And you didn't make just a couple of johnnycakes, like I make for Carrol now. You made a whole griddlefull. Say I was gonna make a dishfull. I'd put in maybe a cup of johnnycake meal for a group. This is for a group. A cup of johnnycake meal, a little salt. I never put sugar in it. Get the water to boilin' and put that in next, and then a little bit of milk to thin it down a little bit. Some of 'em like thin johnnycakes, some like 'em a little thicker. So I used to make 'em thin and then I found out Carrol likes 'em thicker, so I make 'em thicker now. Johnnycake, dried beef gravy, pork gravy, codfish gravy. There was a lot of gravy and potatoes. Always potatoes at every meal too. Mornin', noon, and night, when they come in from the barn from doin' their chores. They'd come in and eat a big breakfast, come in when they had workers with 'em and they worked more then. We'd have johnnycakes and a gravy, like if you cooked a roast you make a gravy with that.

But that was not all she prepared for them, as she suddenly remembers:

Vegetables! I can't leave out the vegetables. Either it was string beans, baked beans, swiss chard, mixed greens. Nettles, dart, plantain leaves – those that's out here in the yard you know, the flat one? You can't use too many plantain leaves though – they're poison. Dandelion, pigweed. My mother used to know what the mustard weed was. We used to make mixed greens, and I've canned mixed greens. Skoke – that's another thing – skoke. I made homemade biscuits all the time. Oh, they loved them too at a meal. Always had to have the nice, warm biscuits. Then I got to making bread, and it was nothin' to make ten loaves of bread. I've got a pail with a handle on it, and I make my bread in that. Well I don't make so much now. I'll make it in a little bowl if I want some buns or bread or somethin' like that. But around Christmas-time, the church always wants the bread for the bake sale, and they always look for me to make the bread. I have to make about a dozen loaves, but that bucket helps.

While dairy farming took up the majority of their time, it wasn't profitable enough to provide for all of their needs. Large vegetable gardens were an essential part of each farm's existence, as Carrol enumerates:

We used to raise up a lot of our own vegetables. Chickens to get the eggs, so we didn't have to buy eggs. Mother used to can up the vegetables for the winter. My wife cans up vegetables here now – ain't too many people do their own canning now. Must have been half, two-thirds of an acre – a big garden. Beets, beans, an awful lot of stuff. Peas, carrots, any of it. Didn't have freezers then – canned everything. Same way in the fall with your fruit. You had a fruit tree had apples on it, or plums, or whatever. Cherries – we used to can them up. We used to raise a big garden. Then if we had some left over, someone didn't have much, we'd give 'em some, or they'd give us some if they had a big surplus – they couldn't use it all. One hand kinda fed the other.

Other foods consumed by the family throughout the year came from animals raised on the farm and fruit trees or berry bushes located in the yard. Betty notes that they still raise a large garden for use at home, although there is one notable change in this year's offerings:

Plate 104: Baling hay. Carrol drives while Robert stacks the bales on the wagon.

This is the first year since I've been here, that he's not planting potatoes. Because last year, oh what a job! They got rotten, even the ones that we bought was rotten, black spots in 'em – in the middle of 'em, all black. We planted potatoes, used to all the time. Well, they always had their new strawberry bed, sweet corn, beans, cucumbers – bread-and-butter pickles we'd make, lettuce, beets, we even canned beet greens and beets – the real beets – big ones, and make beet greens and everything. They've got their asparagus now. Have our own peaches, apples if we get 'em off the tree, we used to have a pear tree, and that used to do pretty well, but that died. Then we had a plum tree too, right over the wall there, and we used to go out and get plums off of that. We used to go out blueberrin' and stuff. Now you can't get nothin' because the birds get it all. Not so many birds 'till they got them starlings in there, and they ruined everything. We got chickens – we've always had chickens and that for our own use, and then we used to have any extra eggs, we used to take 'em down to Gouvin's store and he would take the eggs in exchange for groceries. We did that when the mother was livin' when I first came here. That's where we used to do some of the shop-

pin' and take the eggs and stuff down there, but now you can't. But still we have our own eggs and milk. Do buy the butter, buy the margarine, and I do buy the shortenin'. We used to use lard – take it from the pigs, you know, and melt it up and put it in containers and use that. I even used it in cookies and stuff. Well of course all of that stuff you can't have now. It did come out good, because I used to take that and cut down on the butter, but use about half and half.

Betty notes that they raised and butchered the animals they ate in the past, as they still do today:

We had our own pigs, we pickled our own hams, we had our own salt pork,

Plate 105: The older barn complex, with the milking parlor on the left.

'cause that's what you used to use to make the pork gravy. Take a piece of pork, and it wasn't a little piece of pork – it would be a piece of pork like that – and put it in the greens, or put it in the baked beans or different things like that when you cooked it. Usually they always had about two pigs that they used to have every single year. And then we used to have our own bacon. That old refrigerator they made into a smokehouse and they used to smoke it. Even lately they use it once in a while when they have it. We didn't have beef cattle. Like if a cow got hurt, say, and they knew that they couldn't save her, they would butcher her and then we'd have hamburger, and all the stuff. And then they used to make their own dried beef. After they'd pickle it, they'd hang it on the swing some, and let it dry out, and then you slice it before you package it. I don't do a lot of that stuff anymore. Then I'd freeze the strawberries, make jelly. That's another thing, cookin' the jelly. Grape jelly, blueberry, raspberry – all that.

Carrol describes how beef was handled as well as additional supplements to their diet:

Years ago, father used to kill a cow, hang it up here over in the woodshed, and it would stay frozen almost the whole winter, but you can't do that now. If it didn't get eat up by spring, Mother used to can it. Cook it up and put it in jars and can it up so you could have it durin' the summer. I used to hunt squirrels there in the fall of the year if I had a little time. Half the time I'd go out and husk the field corn and squirrels would be in the field corn. I'd shoot 'em, skin 'em out – Mother would make squirrel pie. Make damn good squirrel pie. They were good. If you got five or six of 'em, they make a good pie.

Things didn't always run smoothly in the meat-processing department, however, as Carrol humorously recalls:

Mike – he always liked to help us kill pigs, when we butchered a pig here. He had a pig, and we helped him butcher it. We was goin' to tip the pig over so we could stick it. He got straddled, that pig went right between his legs and took him right around the coop – Mike right on his back, hangin' on to his ears! That was a

Plate 106: Pump and milk-tank room.

picture. His wife stood there in the door, Frances stood in the door – she told Mike, "I wished I had my camera. I'd like to get a picture of that!" He was goin' around, pig poundin' right up and down and he hangin' right on to his ears. Yeah, we used to have some good times with Mike.

Early on, before electricity came to the farm in the 1930's and long before the advent of reliable long-term refrigeration, the milk produced by the cows was stored in their old milk house in large cans sitting in cold water. For household needs, canning provided the means of storing food for long periods of time. When asked about the scope of her canning operation, Betty replies:

Oh golly, it depended on how much stuff we got. Like the first year that I was here, I went out with the father and we picked blueberries and stuff – I bet you we canned – well the mother was helpin' then too, and I bet you we canned almost two hundred quarts of blueberries. Use it for pies, use it on the table for preserve, because they like that preserve at a meal to go over the biscuits. And then we used to do a lot of, like I said, the greens. Go get cowslips. This year I got fourteen quarts of cowslips. I've already canned some rhubarb-strawberry jam. I've got eight jars of that. And then you have your peaches. And he got so they liked peaches frozen better, so I just slice them up and I put that Fruit-Fresh on 'em and the sugar and put 'em in a package – little individual packages. We even canned meat – used to. Like if we got deer. We used to take and cut it up in small pieces and pack it in the jar, but you couldn't pack it tight, so we always used to take a knife and stick down in the middle of that and wiggle it a little bit. But we put the salt in it and just seal it up – no water – nothin'. It made its own juice, the meat did. We've canned that, we've canned meat, we've made mincemeat. Mincemeat is nothin' but meat ground up, raisins, apples, sugar – that's about it I guess. The meat was ground up or cut small. You had it fine. Used to make watermelon-rind pickle, green-tomato pickle, green-tomato preserve. Canned the peaches too, when you had too many. Between getting meals and trying to can stuff I was busy, but you've got to do it when the stuff is ready. You can't say, "Well I'll leave it for a couple of days." It rots on you. You've got to do it while it's ready. It keeps you busy all the time. But then I used to get the children to play a game with 'em, you know, and they used to help out quite a little bit, like picking the stuff and getting stuff from the garden and everything. And they used to help out. But then of course they had their school work and stuff like that. They used to take those pails of milk and that, carry them over and dump 'em. They used to work hard too.

And as with any good cook, the demand for their work never ceases. Betty recalls what happened when she baked a little less than she once had:

I did all of the cookin' when the mother got sick. Well I did some of the cookin' too before that. I used to make bread and make cookies. I never forgot it – the father, after the mother died, I didn't make the cookies as I used to. I made 'em, but I didn't make 'em

Plate 107: Old Farmall tractor and loaded haywagon.

as often – I didn't keep the cookie jar full. And he says, "I don't know what's the matter, but that cookie jar isn't full anymore." I don't do much of the sweets anymore, account of Carrol. Orrin likes his sweets though.

Beyond the cooking and chores around the farm proper, Betty somehow also found time for needlework. Much of this work was also directed toward practical needs:

I made the children's clothes – used to. Make 'em sweaters. I got a sweater in the other room there that I made for Carrol with a cow on the back and two cow heads on the front. Orrin's got a sweater when he was a little boy, the same thing, but I had to cut it way down, the needles and the yarn and stuff – fine yarn. I made him one. I used to make the girls all their sweaters. I use to make 'em dresses. I used to love to do my needle work and stuff. Knittin' and crochetin'. Hooked rugs. Marilyn's got a rug in her livin' room that I had hooked for her ever since she's got married. I made them when they got married and a quilt with the blocks embroidered. I did it lots of times when I used to put the children down for their naps.

Betty fondly recalls putting Dulcie down for her naps in the cold winter months:

Dulcie, and that, when she was little, she'd say to me, in the winter time, "Momma, put a popping stick in the wood stove," when she went to go in to take her nap. You'd put a stick

Plate 109: Robert welds new steel side panels for the truck.

Plate 108: The cow lot across the road.

in the stove and it would snap, you know, and she liked to hear it snap. I'd take her in the other room and that to take her nap and that and she laid down, and you could tell she was listenin'. All of a sudden she was fast asleep. Poppin' wood – that's what she called it.

And as with all parents, she had fun stirring up the children's imaginations – something she recently discovered is also passed down to the next generation:

Another thing too that I got with Dulcie. She was little and I had her over here lookin' out the window and she says, "That's a hill there mama – a big hill!" I said, "Yes – that's a dinosaurs been buried there – that's his back." Well she believed me. For years she believed me that a dinosaur was buried up there. I didn't realize she was going to do that. She brought it up not too long ago and told her children!

Carrol too remembers teasing the children when they were young:

Christmas. I took a cow out here in the barn, I took some of the hair from off her neck – the hair was nice and white off the cow's neck. I was goin' to take it off the tail, but the tail was dirty. I took a rat trap, where they put the food for 'em, and I put the white hair in the trap, and I said, "I caught Santy Claus' whiskers! That'll learn 'em!" Oh – they was mad at me! They was tellin' everybody, "Grandpa caught Santy Claus' whiskers." It was a long time before they ever found it out. Yeah, I used to have a good time, especially with Marilyn and Helen. I could get them goin'. She found it not too long ago and showed it to Alex. "See where Grandpa caught Santy Claus' whiskers? Old, mean Grandpa!"

Daily work on the farm not only produced lots of food, but lots of laundry as well – another task that kept Betty busy from dawn 'til dusk. This also brought back pleasant memories for Betty, beyond the added labor:

We had a wood furnace when I first came here. I used to dry the clothes over

Plate 110: Stalls in the old section of the barn.

that. The father had a rack he made over that. Marilyn still yet hasn't forgot it. I used to have to hang the clothes out – didn't have a washer and dryer then – only the old wringer type and stuff like that. Used to bring it out here and wash clothes, take 'em out, hang 'em on the line. Freeze – then wonder why something's wrong with my hands. But anyhow, bring 'em in, the men's strip-ped overhauls and their dungarees and their pants and stuff would be stiff! It looked like you had them in your hand. Put 'em over that rack and you'd hear it go down on top of the furnace when it thawed up – dink! dink! The children used to love it. The diapers even. I used to have to hang the diapers out. The only time when we got the washer and dryer was when Orrin was born, and then I was goin' to have Mildred. Orrin wasn't even a year old. She was born the next year, a few days before he was goin' to be a year old. He said that where I was gonna have two so close together, you know, and Carrol was only gonna get the washer, and he says, "No, get the dryer to go with it." That's how come we got the washer and dryer and had 'em.

North Stonington was also known as Milltown for the number of mills once occupying its many streams. Where as today we can hardly guess where our grains might come from, Carrol and his father would go to get them ground locally:

Used to call it Milltown. Father tell mother, "I'm goin down to Milltown." They had a sawmill up in back here. Steam-powered, although I guess they did have a water-powered one. They had one over Clark's Falls, was a mill over there, besides the grist mill. Used to be a mill over there by Babcock Road. Some type of mill, and then there used to be the grist mill down there. Father used to tell about. Some of 'em take corn down there and get it ground for meal. Used to have the old corncrib for field corn. Carrol Maine over there used to raise field corn. A lot of people used to raise field corn. Coats over there used to raise it – the old folks used to have a corncrib over there. Uncle Palmer down there. A lot of these old places had a corncrib. When Clark was runnin' the grist mill over there to Clark's Falls, Clark used to run it, and I've been over there. I remember one or two ton of corn sittin' on the floor there to be ground. The old mill would be runnin' steady. Now they don't run it at all. I think they used to grind wheat. I don't know if they ground oats or not, but I heard 'em tell

Plate 111: Charlie Cole repairs a flat tire on the tractor as Orrin looks on.

about grindin' wheat.

Today, most of the products needed in the dairy business are delivered to the farm in bulk by grain or trucking companies. Thirty years ago, it was often up to each farmer to get his materials to the farm in whatever manner they could. Carrol remembers his father's adventures in moving commodities from one place to the next:

Father used to make five or six cans of milk – forty-quart cans. Then they'd come pick it up – Albrow come pick it up, haul it to Providence. I think they was getting around, between three and five, six dollars a hundred pounds. Cripes, you could buy a pair of shoes for a dollar. Buy five gallons of gas for a dollar. Buy a ton of grain for twenty dollars. Father would take that old Buick and go down and pick it

up. He'd go down, take the back seat out, put it in there. You wouldn't put a ton of grain in a car now! He'd have it packed right up full, that old Buick. Once in a while a tire would blow out. If it wasn't stormin' goin' to town then he'd take the old horses, team wagon or dump cart, go down and get the grain with that. Sometimes they used to take wood down to Westerly – load of wood on the old team wagon – deliver it and then bring back grain or bring back somethin' else. One time he brought back a barrel of molasses he told about. An old wooden barrel. Heard him tell about drivin' turkeys. Carrol's over here, it was the fall of the year, drive 'em clear to Providence. Flock of turkeys, get a bunch of guys, drive 'em to get 'em killed down there. You know that's quite a little step. They used to tell about some of the turkeys would get pretty sore feet, of course all gravel roads. They start out in the mornin' and they said they'd make it there by night.

After Carrol's father passed away, he and his brother George ran the farm together and raised their children. Orrin, Carrol's son, and Robert, one of George and Frances' two sons, grew up on the farm and stayed on as adults, choosing farm life to careers in other areas as had their brother and sisters. For Robert, being raised in the middle of the farm was heaven and certainly shaped what he would do as he grew up:

I remember vaguely when they built the free-stall barn, which they built in '68. Ever since I was six, seven years old, I was always up at the barn. I didn't like the house – I wanted to be outside. Now my brother Maurice, he did chores and stuff, but he didn't enjoy the farm work as much. Farming has been pretty much everything to me. Even when I was a kid, if I played on the floor with toys, it was always farm related, it seemed like. I always wanted to do that.

While Robert shies away from taking credit, the operation of the farm is largely in his hands as Carrol is now in his early eighties. The operation of any dairy farm requires a large and varied skill set. In today's society, most people focus on one fairly narrow area of expertise and hire others to cover what they cannot do themselves. Farmers, either by choice or necessity, must learn to do a thousand different tasks, and Robert is no exception. When asked how he came by the know-how for all of his particular skills, he had this to say:

My father used to feed the cows. I went up to help my father do chores, and gradually he got me to feed the cows. Orrin, at the time, didn't care for the milking, so after my brother left the farm I ended up going into milk. As to breeding the cows, my uncle asked me one time if I wanted to learn how to breed cows. The breeder we had coming in was going to go up on fees or something, and I said, "Yeah – I want to." And about that time I was starting to get more into the veterinary part of it. I learned some from my father, I learned some from a neighbor named Ken Lattimore, and between my father, Ken Lattimore, and a veterinarian by the name of Bruce Sherman – he's since retired and works for the state now – I really liked doing the veterinary work and whatnot – so that's how I took on that job. So whatever I kind of came to I did. I enjoy mechanics. I don't know, you talk to different ones they say I do it all up here – I don't know. Jack of all trades, master of none.

While he may not admit to it personally, one only has to follow him for a day to see that he is much more than a jack-of-all-trades as he works the farm. As with Cool Breeze Farm, Robert and Carrol do not hire permanent additional help onto the farm, which keeps Robert hopping throughout the year. He had this to say when asked about his role on the farm:

Plate 113: Grasses maturing in the summer in front of the barn.

Plate 112: Orrin attaches the mower.

The planting I do pretty much myself; the first spraying I will do myself. If it needs to be sprayed again, that sometimes depends on whether I have time or not. Sometimes we've had to hire somebody else to come in. For the mowing and haying, between my uncle and myself, we do the mowing, and I'll end up doing a lot of the haying. If I need some help, I'll get some friends or whatever, or I trade off with Lewises. Then when it comes chopping time, usually I have friends come in, or trade off or whatever else. And I do help out a lot of my friends, but it's all a trade-off. The milking, Carrol and I mainly do right now, although I'm in and out of there sometimes depending on what else needs to be done. I would like to hire another young kid, but

we'll see what turns up on that one.

The waste products from the cows are spread onto the fields along with commercial fertilizers when needed. They have also used chicken manure in the past, but Robert feels that may have brought on problems of its own:

We put down some manure, some fertilizers. I don't have enough manure to do it all, and so I use the fertilizer. I know I could get chicken manure, but years ago there was a big theory that with the chicken manure there was more weed seeds that came in, and some of these were hard-to-control weeds, so I kind of stayed away from that. There's this type, this weed, called burr cucumber. It's a mean thing, and a lot of the guys that had it, that's where they claim it came in from – was the chicken manure, so I try to stay away from it. I've got a couple of weeds up there that I'd like to get rid of – I'm trying,

Plate 114: Carrol leaning against the horse-drawn hay rake he once used.

but it's not totally happened yet. The grasses get manure, and then for second cutting I'll come in with a commercial fertilizer, just so it can give us a better crop of second cutting, which is more of the grass you want to feed the cows to make the milk with a higher protein and everything – with tender shoots, where the horses seem to like more of the coarser feed. With the first cutting, if you mow it early enough, they claim it's just as good as the second, but usually I'm planting and I don't get to that at that time. We usually get about three cuttings. On the rarity, maybe there would be one or two lots you could get a fourth, but typically three.

Plate 115: Stainless bucket in sink.

As with all of the farms in town, the size of the herd they milk has changed over the course of years. Starting out in the twenty-five to thirty cow range when Carrol was a boy, it expanded to milking more than three times that before Robert's father died:

When my father died, we were somewhere around ninety cows – milking around ninety cows. The most I think we've gotten up to was a hundred and ten, a hundred and fifteen, but that was back before my brother left. When you have that many people it's not quite so bad, but a few years ago, I was getting run pretty ragged. So we got down to more like seventy in range, and that was more comfortable. I was doing better. And then we weeded out some of the herd, and now we're down to about fifty. We're milking about fifty. We're probably around, between the beefers I've got, probably around a hundred head or so. A hundred and ten, something like that, total, on the farm.

Modernization is key to helping Robert keep ahead of the work that must be done each year, but he admits they have modernized a little more slowly than he might like:

There are certain things I have been able to change – like the no-till planting. My father was never no-till, and Carrol was pretty hesitant about it in the beginning, but he has since changed his tune and whatnot on that. There's certain other things. One of the things I can tell you that surprised the dickens out of me once – we were having a very wet year – and well, my father was still alive at the time, and we weren't able to get the hay off. It was a year similar to last year. My uncle said something to me about had I ever thought about getting a round baler, because he and I were milking at the time, and I said, "Yeah, I want one, but we can't afford it." And I was trying to go into it gradual. "Well, let me talk to your father." So he talked to him there, and they came to an agreement, and here it is in June and I had to go find haying equipment. They turned it over to me to figure out what equipment, and so that's when we bought the round baler. That

Plate 116: Orrin and Carrol attach a fertilizer spreader to the International tractor.

year I just bought the round baler, and I borrowed Eddie Bill's wrapper. That was after I had Sherwood Damon come a couple of years and bale hay for us and wrap it. And then I bought the wrapper in the fall of the year, and then after Sherwood Damon died, that's when I bought the tractor with the grabber. So it came very close to one lump sum, but once you figure those three pieces of equipment in there, you're talking roughly about, well, the baler was twelve thousand, the wrapper was seventy-five hundred, and the grabber – just the grabber that goes on the front of the tractor – that was three thousand, and I got a deal on that because I bought it through Stanton. At the time, Niles needed one, and Lewises needed one, so I bought three together, and they each bought one. The tractor I got fairly cheap. Still, it's a considerable investment, and

parts ain't cheap. Last year wasn't so good for selling hay because the economy is so bad. A lot of people that might have called have gotten rid of animals.

To feed their herd today, Robert says they plant more than they need:

Right now, about seventy acres of corn, and about sixty acres in grass. I've got corn that I chopped last year I haven't even touched. Years ago they would go through all their corn, but of course we had more animals and whatnot. In the course of a couple of weeks, we'll probably go through four-and-a-half to five ton of grain, and that's just to supplement the feed that we've got.

Like everyone else, Robert must wade through the dozens of varieties of corn now available to him, and he faces the same troubles the rapid changes in the market bring. He additionally faces the same problem Cool Breeze Farm faces when purchasing supplies for his particular farm – an economy of scale. Without the purchasing power of the very large farms out West, costs to him are elevated:

I pick out the corn. I have one seed company I care for – I've had a couple of bad experiences with other ones – and that's just not getting the yields I think I should get using it. They have field trials every year – you try to look at that, but there are so many seeds out there. I also try to listen to the salesman somewhat. I tell him, "This is what I want, this is what I want, and this is what I want." And then well, he'll give me, "This is kind of in your category," and then I go pick it out. I try to narrow it down somewhat. For years my father used Agway seeds. Then I got to picking out corn, and I tried a couple other different companies. Mainly I'm with one company because I buy twenty bags of seed corn and that gives me a bigger discount. When my father was alive, we'd do it with a couple different companies, but as with any business – bigger the volume, bigger the discount.

Much as he would like, he can't always contain costs by purchasing more at once for later use:

You can't buy more and store it, because the germination sometimes goes down. You also have your rodent population if they're going to get into the seed corn. Say we had a bag left over, you've carried it for a year, and yeah it comes up alright, but you wouldn't want to bet the bank on it. And maybe it's one of them varieties you didn't like that you had. Of course it seems like most of the time I go in and one of the varieties I used last year is not available anymore. What I end up doing is sometimes I plant two varieties in the same field. I'm able to see what one variety can do against the other. I've got a four-row planter, so three rows of one variety and one row of another type, so I can tell the difference. Years ago, even when my father was alive, when we had two different seed companies, we'd do that all the time. You'd try to get the same day length too though, so they'd be ready to chop at the same time.

Plate 118: Clouds building above the fields and barns.

Plate 117: Old grain shovels.

In addition to the corn fed to the animals, they also need to harvest the grasses necessary for balancing the cows' nutritional requirements. According to Robert:

We use mainly mixed grasses. Orchard grass, because a lot of our fields haven't been turned over and changed. Slowly I'm trying to do that. The lot behind that hill out behind the barn, that was always corn. I never remember that being hay. They tell me it was at one time. Some of the hay lots, I've heard my father and uncle tell about them being corn lots – I've never seen it. It should be done more, but I only have so much time.

Robert says they mix their herd's feed largely by eye at this point in time, gauging their success by direct observation:

A lot of people do it by weights. I don't have the facilities for that yet. That would be the better way to do it. Mainly I do it by eye, or I have a pail or a shovel, and this is roughly what they usually get, or a scoop. I have a measuring device. It's not as scientific as some. I try to watch the manure, see what the consistency is, how their coats look, how they're acting, how the production is. Within a couple day's time, if something's drastically wrong with the feed, if they're still eating it, you'll drop on production or whatever else. If you know there's a difference, generally even in twelve to twenty-four hours you can start to notice a difference.

There is a delicate balance that must be kept in the milking cow's diet in order to maximize the production of milk. Robert often points to the similarities between cows and humans:

We put in bi-carb for a buffer – a rumen buffer. It keeps the acid down in the stomach – keeps the ph level. Very similar to us. Say for some reason the feed's more acidic or they get a bellyache. Usually your stomach, when you get a stomachache, your ph in your system has changed for some reason. This is just to keep them on an even keel. Cows are very similar to humans. They're all

different – there are different characteristics, different demeanors – they're all different. I mean I've had some that are very wild, I've had some that are very docile.

A cow's health is, of course, vitally important to the farm, and while house calls are still required due to the size of the patient, veterinarians have also changed the way they practice over time. Carrol recalls a health problem coming in with an animal his father purchased from Carrol Maine:

You could always get a vet to come in and doctor a cow. Old Dr. Heath down there in Stonington, he used to come up and doctor a cow. Cripes, Carrol bought some cows. Father bought a cow from him and she had a distemper. Top of the whole herd, Carrol lost, I don't know, four or five cows. Father lost two or three from buying that cow 'cause she had shipping fever. Dr. Heath came up here, and he finally found out what they had. Took the lungs out of one of the cows that died, and sent it off, I don't know, up to UCONN or where he sent it to. He doctored them – he stayed here all night. Go from here, over to Carrol's and back here – see how that cow was doin'. You wouldn't get a vet stay with a cow all night now. Slept out here in the barn. Father tried get him to come in the house here, "Nope. I'm gonna stay out here in the barn." He 'nocculated all the cows here and 'nocculated them all over to Carrol's. Old Dr. Heath – that's where Ken Lattimore got a lot of his trainin' from, Dr. Heath. He used to travel with him and help him out some. He never went to school for it, and he was better than some of the vets today. He used to doctor Niles' cows over there. A cow was somethin' like two or three hundred dollars. It was a lot of money.

Robert's interest in his animal's health came from working with his father, a proficient farmer just down the hill, and with their visiting veterinarian. While there is a limit to what farmers can do in certain situations, their broad knowledge of their animal's health gets them through most medical emergencies. Not only does having the ability to treat their own animals save them on vet bills, but it also keeps the herd producing milk. There are a number of common problems affecting the dairy herd, as Robert enumerates:

Mastitis mainly is coming from the lactating animals. That's

Plate 119: Robert moves a thousand pound round bale of hay to the wrapper before storage.

an infection that's introduced in the udder. Twisted stomachs. That mainly comes from the cows when they've just calved. And if for some reason they go off feed, which there's a variety of reasons for – their stomach fills up with gas and it inverts. And there's two ways it could twist. Usually the right side is the more lethal way – it will cut off the blood flow. The left side, it could twist, and those are the ones that it's not as quickly that you have to remedy the situation. They can go longer, because you're not cutting off the blood flow to where it comes out of the stomach, going into the intestines. You also have ketosis, which is a form of low sugar. Usually when they get ketosis there's a chance they can get twisted because they're going off feed. That comes from a fatty liver, a cow being over-weight. Then you have your milk fever, and that's a calcium deficiency right at the time the calves are born, as they put so much calcium from their system into getting the calf out from muscle contractions and making milk that they get low on it. That's not a definite thing every time. Cows have been known to get pneumonia. Sometimes they've been known to get hardware, which is metal in their stomachs. Any animal could get that, but that's mainly because of mechanized feed and you could have a bolt come off or nail or something that will puncture the stomach lining. The way our cows are grouped, I want to say there's more fretting during the calving time – it's a more stressful time. That's when things are more apt to go wrong, other than the mastitis. You try to keep where they are laying clean – try to prevent that infection. And every situation is different – what works on one farm might not work on another one, so that's why there's a lot of different teat dips out there and whatnot. We've got one, and my theory is, if it ain't broke don't fix it, so we're still using that. When we start having problems, then we'll change. We've had instances where I didn't feel the antibiotics were working. We didn't know if the mastitis germ that we had was starting to build up a resistance, or we had a different variety in there, and so I've had it tested.

A cow's general well-being is often judged by its stance. Dairymen can read much about the cow's physical state in the condition of its feet and legs. Modern dairy cows are somewhat prone to problems with their hoofs and legs as they remain on hard surfaces longer than their ancestors once did. They were, of course, made to wander in fields on soft earth rather than in free-stall barns. Besides careful trimming of their hooves to prevent problems, Robert also must de-horn his animals, for obvious reasons:

All the Holsteins we've got, we have to de-horn all of them. Years ago, they were left for show and everything else. But a cow, when they get to head-butting and everything else, they can drive a horn right into the side, or rip flanks. Very similar to deer.

His interest in animals also extends to the very important breeding program to which every farm must attend.

Plate 120: Carrol rakes the dried hay into windrows so it can be baled for storage.

Plate 121: Cows seem willing to pose for photos.

As with all of the farms in town, the Miners artificially inseminate when they can, using their Hereford as a clean-up bull. As Robert prefers Holsteins for his milking herd, the crossbreeding with the Hereford produces calves that will be raised as beef animals or sold:

The reason we got that Hereford bull is, the heifers run loose across the road, so we just turn the bull loose across the road. Sometimes if you can't get them bred artificially, then sometimes he has better luck. You use him kind of as a clean-up bull. That's why he's a beef breed. We're using the Holstein for semen artificially and the clean-up bull is him. Mainly I'm not milking those – I get rid of those animals or I raise them for beef.

I've only got one white-face that's in that milking herd, and that was going to be a beefer. But that came down to I had a bull that was about the oldest bull the Lewises ever saw. My father bought it. Dave Lewis had picked it up and my father saw it and told him, "When you go out to auction there, if he don't bring too much bring him back." And he brought him back and he had to be some twenty-five years old, and gentle. It got so I really felt I could do anything I needed to with him. I could put him down in the pasture in the morning, and I'd go down there later, and he would want to stay with the heifers. You'd get around him and give a little tap on his rump – he'd come right up out of the pasture. If that gate was open, he'd go out that gate, go right up the driveway and right into his stanchion in the barn. He was that good. He was probably the best, calmest bull we ever had. In his day he probably weighed a ton.

With everyone who works so closely with animals, there is a natural interest in all breeds, and Robert is no exception:

I prefer the Holsteins, Well, that's what I always grew up with. I've seen other breeds, I've read about and talked with other people about other breeds, and I've always said I want to have one milking short-horn. I don't know why – I think they're kind of a neat looking animal – but so far I haven't gotten one yet, and I'm not in any big rush to go find one neither.

Robert jokingly transferred his abilities with the birthing of calves to the birth of his son Jason:

When my wife Cheree was pregnant with my son, and we had to go to birthing classes for Lamaze, that was the most fun I've had. She hated it, but it was the most fun I've had. The teacher would be giving the course, and I'd be shaking my head "no" or "yes" and she'd look at me. She said, "Well, you've seen cows give birth." "Oh yes." And I would proceed to tell her what we did with cows. When they were talking about the feeding class, I was waiting for that class. I got one of my big calf bottles and I had her put it in the bag. She's going on about how you have to feed a newborn baby every three or four hours, and I'm just there shaking my

Plate 122: Breeding chart from Federated Genetics.

head no, no no. So she said, "Well, what do you say?" I pulled out this big bottle and I said, "One of these in the morning and one of these at night – good for the whole day!" Now probably half the other people were mortified that were in that class. One of her final comments to Cheree was, "When you start having a baby – if he starts coming at you with the chains – RUN!"

With milk prices fluctuating and costs continuing to soar, Robert has also thought about diversifying, although it is difficult for their farm with so little labor available. The daily chores are nearly as much as they can handle at the moment, but it certainly occupies his thoughts:

I mean I've got ideas in my head – what can be done. Right now, at least the way it appears, you can't make it really in farming with just one product, so I do agree with diversification. It's just trying to figure out where I'm going to get the money to be diversified. I've got some beef. Mainly we used to raise them just for ourselves. Since my father died and we have less animals we've got more

feed. I mean my father and uncle very rarely sold hay or whatever, and I'm selling hay now, so I guess I've diversified somewhat already. I'm raising more beef because I've got the feed, so I may as well feed it to something. With the beef, it takes a little less care for the beef, you know. I'm not worrying about milking and everything else.

Plate 124: Feedbags are no longer reused.

Plate 123: Carrol and Orrin move fertilizer with the help of a loader.

With the economy the way it has been for local farms, even branching into the raising of beef cattle can be uncertain. As Carrol sees it, however, farmers by nature must also be risk takers:

Calves ain't worth nothin'. Robert got some white-face out there that he's raisin' – Herefords – hoping that he can make a head of 'em when it comes time to get rid of 'em. You take a calf now, by the time you send it to auction and pay the commission and pay for truckin' to get it up there, well, Esther sold some and only got two dollars for 'em. You don't get ahead very fast. He figures he can give 'em the milk and raise 'em up – they might be worth somethin'. It's a gamble you gotta take. You have to gamble. They tell about goin' up to the casino – farmers have to gamble too. Just like goin' out there mowin' down a piece of hay – you gamble on it whether you're gonna get it wet before you get it in or whether you're gonna get it dry. Same way with corn, when you're plantin' corn. You don't know whether it's gonna come off a wind storm and blow it down, lose some of it, or what you're gonna get. Get them damn wild pigs around here and probably we'll get nothin'!

While Carrol may now be worrying about the introduction of wild pigs into the area and the damage they could do to his crops, he also notes those animals that are no longer found around the farm:

Used to see pheasants, quail. You don't see them now. Used to hear the quail in the springtime, hear 'em holler "bob-white." I ain't heard a quail in years. Another one we used to hear in the springtime was whippoorwill. If you heard them, you could plant anything then – you didn't have to worry about frost when they holler. But you don't hear them now. Last one I hear holler was when Carrol was livin' there and they used to be around the pond, in the brush over there. You'd hear them holler. Or sometimes they'd be down here on this pond. They say it's time to plant corn when you hear them holler.

Never saw a possum – never heard of 'em. Skunks – had a lot of skunks. Wilkinsons down here used to go skunk huntin'. Then they'd come up here to school the next day and the teacher would put 'em way over on one side. They were pretty ripe! There used to be bears around. There was a bear's den up here in back of Wyassup.

He also talks of other natural signs that helped them with planting and harvesting:

You stand out here in front of the house, you'd see the boats right on the water – sailboats. The white sails. Most anyway you saw it, when the ocean was clear, it would storm the next day.

Plate 125: Robert moves the corn chopper from the field he just finished into the next.

Make up your mind, being so clear, next day you'd have a storm. Rain, or in the wintertime, although of course you didn't see the sailboats, you'd get a snow storm. When you see the mare's tails in the sky, them white clouds there, made up your mind rain's a comin'. When you see the smoke come out of the chimney and go down to the ground, that's another indication it's goin' to storm. Clear away at nighttime, will stay clear forty-eight hours. Yeah, I can predict the weather a little.

While the weather may still be predicted through observation of the distant ocean or a local chimney, many other things have changed throughout Carrol's long life in the dairy business:

They used to take the old milk pail and rinse it out and hang it over a bar post. You didn't hear of any inspection. I don't know when they did really come in, but I seen them take the old strainer, wash it out, hang it over a bar post or something out in the sun. Today that wouldn't be sanitary. There's a lot of inspections. Milk production, and well, your cows gotta be clean. If you ain't got no problems they come every six months, but you never know when they're commin'. If you got problems, they're right here to see what's goin' on, where your trouble is. If you got high bacteria or anything they're right there.

I test the milk here for antibiotics. If we treat a cow, I test it before it goes in the tank – make sure the antibiotics is cleared up. If the dairy finds it, you're in hot water. You have to pay for that load of milk.

The milk they produce is handled by DMS, and according to Carrol is picked up roughly every other day:

Plate 126: Carrol compacts the newly-chopped corn silage with a tractor.

DMS – Dairy Marketing Service – is the one that pays us for it. Now they take it up to Garelick, or they take it for cheese down Providence, or they can take it to Springfield to Agri-Mark if Garelick is loaded up with milk or something. It may even go to New Haven to a cheese place out there, or it could go to Guida's. So you never know. Right now we make about two thousand pounds every day, about every four thousand they pick up. Sometimes it's over – not right now – this time of year. Right now we're way down. Got a lot of dry cows out there – gonna calve. We had trouble getting them bred there, some of them. Supposed to calve in April.

At this moment in time, Robert's son Jason is much too young to know if he will have the same interest as his father in running a farm, and perhaps only time will tell. Running the farm is certainly in Robert's blood and he hopes his son will at least understand his passion for it as he grows older:

I can tell you at his age I was more into it than he is, but of course, I grew up right up on the farm, and really he hasn't. This year he seems to be taking more of an

interest. I hope he realizes when he gets old enough why I did what I did. I know I haven't spent as much time with him as a lot of fathers do, but I hope he understands and how much I love my job, which many people can't say they do. I can't say I get up every morning wanting to go to the farm, but when I go it's always different – there's something different every day. When I go in the barn in the morning, usually my plans get changed two or three times easy by the time milking's done. I can get up with a focus in the morning – usually it gets changed.

While none of the farmers in town have been willing to predict what the future may bring to the dairy industry, Robert knows that the farm on the hill will remain:

It's going to stay open. I can't tell you how, but it will. There wasn't much open land when my grandfather bought it. I think he bought about a hundred and some acres, and since then they've added to it and we've added to it. I was involved in some of the clearing of the land, and it would really hurt my eyes to see it developed. I can't say that it won't ever, but I don't want to see it developed. If you go to the next generation, even if it's not a working farm, even to grow back up with brush, I hope they understand what it took to get the land where it was and why we don't want to see it developed.

Plate 127: Three generations: Frances, Jason, Cheree and Robert Miner.

**Scenes from the
George & Carrol Miner Farm**

The Palmer Farm

Plate 128: View of the Palmer Farm from across a cornfield. The Palmers moved onto the farm in 1897.

The Palmer Farm is found in what is known as the Clark's Falls section of town. Located at the intersection of Clark's Falls Road and Dennison Hill, the farm is owned and operated by two brothers: George and John Palmer. With a herd of around five-hundred animals, they currently milk about two-hundred and fifty cows. The balance is made up in calves and heifers under two years in age. This makes them the second largest farm in town as measured by milk production. Early on, the farm belonged to the Clarks. It was leased at first by the Collins family (George and John's great-grandparents) until they eventually bought the property. In turn, George and John's grandparents bought the farm from them. According to George:

They moved on the place in the late eighteen hundreds. There's a lease in the house, from 1897 or something like that. I think they bought the place, this is the great-grandparents, about 1917. They were Collins – her name – whatever that meant, and she up and died about three years later – not an old woman – and then my grandmother and grandfather bought the place. It became my father's in the early seventies, and John's and mine in the mid-nineties. It's been added onto – there's a piece over here they bought in the mid-fifties, and then I own a piece that is contiguous down here that I bought in '94. And then John and I bought a couple of things together, and he bought a couple of things by himself – well, him and his wife Patricia.

As John states, both the Collins and the Palmer sides of his family were farmers, with the Collins side in the area and the Palmers coming from Voluntown, the next town over. When they began working the farm, there were many fewer animals than they have now, and, as John says, everything else was much different too:

You know, back then, they probably had to milk them by hand. I couldn't tell you when they started using a vacuum pump and milking them into a pail. Of course back then they didn't have the refrigeration like we have today. They'd put it in a forty-quart can and set it into a spring somewhere. Had to have a box made with cold water in it, and they'd set the can in the water, and that's how they kept it cold. They were self-sufficient. My grandmother was a school teacher too. I couldn't tell you when she started, or when she retired, but she had only three children. Kind of a small family for back then, but they made a go at it. My father was the only boy, and there were two girls. My father and grandfather farmed together, and of course my grandfather, he farmed up until he just couldn't do it no more – he got too lame and he just couldn't do it. My dad did have some help, and George came along. He is older than I am and he helped and just kept going – that's all. George is my only brother; I've got three sisters.

Raised literally in the middle of the farm itself, both John and George had chores to do around the farm to help their parents. John recalls some of the earliest things he had to do:

I can't remember what my first chore, or responsibility, was – probably taking care of some young heifers or calves. When I was a kid, I had a bicycle in the back of the house. I'd put a five-gallon pail of grain on the handlebars and ride up the road and feed the young heifers.

Although his three sisters never took the interest he and his brother did in farming, John assumes

Plate 130: John Palmer as he gathers bales of straw.

they would have been called upon, like everyone else, to help on certain occasions:

When they were young, they probably helped out some with haying. I'm sure they learned how to drive at the farm, the old, old truck, or tractor, or something like that when they were haying, but I can't remember them milking cows. I don't know why they never got interested in farming, but they just didn't. I'm sure if they got asked to head off some cattle, they had a cattle drive going on, they took part in that. The youngest, Joyce, she used to help in the garden. She liked getting her fingers in the dirt and planting flowers. I'm sure they don't regret growing up on a farm.

While John was still a kid, his father and grandfather worked out of two old barns across the road from the farmhouse before expanding into the gambrel-roofed

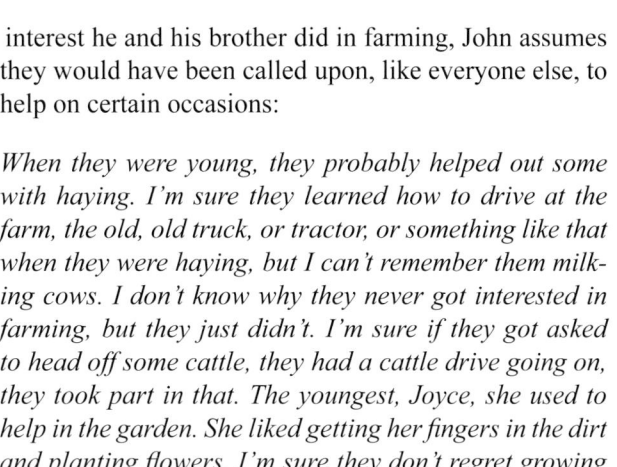

Plate 129: An old cow bell.

barn still in use today:

Back then, when I was just a kid, my dad milked in the two old barns across the road. The two barns held thirty-eight between them. The bigger barn there, the one that has the gambrel roof on it, there were sixty stanchions there. We probably had thirty-eight or forty cows when we moved into that barn, and that was about '63 I believe. That barn held sixty. We milked cows there until about '85, and put the milking parlor in. We carried the pail and the machine from cow to cow and had what we called a dumping station. It's a piece of equipment that you wheeled down the barn. It had an electric motor on

Plate 131: Chris Buck and Becky Loignon.

it, a pump, and a stainless-steel jar that you pour the milk into, and it would pump the milk to the tank. It saved a lot of walking – that's what it did. It had a good-sized rubber hose on it, and you'd just string that along the ceiling, and it pumped the milk to the tank. The cows went outside then – they didn't stay in the barn 24/7.

As with the brothers on the Beriah Lewis Farm, George and John decided to go into the dairy business with their father and expand the operation. Having met the Amish and Mennonite builders who constructed the free-stall barn for the Lewis farm, they too decided to modernize with the same type of building. As John relates:

The free-stall barn is only twenty years old. They've been around for years. Probably early sixties – they've been around at least that long, maybe longer – I'm not sure. Probably the fifties. It's called a free-stall barn because the cows aren't tied up or held in one particular spot. They can lay down in a stall or get up, leave, go eat, come back, and lay down anywhere they want. The barn was built by Mennonites and Amish. The Lewises had one built the year before we did, and that's how we met those people. There is a slatted floor, so the manure just falls into a pit, like a cellar in a house, and it just stays right there, and there's some clean-outs along the side where you hydraulically set a pump down into the manure, into the pit, and you can agitate it or pump it.

This enormous structure has stalls for some 262 animals. Because cows left outside throughout the year can withstand New England winters, the walls of the barn are constructed with an opening between the lower wall and the roof. Curtains can be raised or lowered over this space to ventilate the building and keep the animals either cool or warm. The airy nature of the structure helps keep health problems to a minimum. Because the cows are free to wander around at will, there is less stress on the animals, which in turn leads to a healthier and happier herd. This type of barn also saves many man-hours because of its partially self-cleaning

nature, allowing George and John to concentrate their efforts in other areas. As John points out:

Inside the barn, you don't have to scrape it – there's no handwork as far as getting the manure to go through the holes. You just clean the stalls out a little bit – some of the cows turn around in the stalls and they mess in the front. It's a little more work that way – it wastes more sawdust, but the majority of them go in the alley. The more cows you have, the better it works, because they step on the manure – that's what makes it fall through the holes. Of course, a dairy cow is fed a high

Plate 132: George Palmer in the milking parlor.

protein diet and it's quite soft anyway. Softer manure than a beef cow, by all means. You can have more of them with less labor. You might have more feet problems, because the feet might be a little damper than on cows that go out in a pasture. You know that when you have a stanchion barn, usually you have a smaller herd anyway, so your cows will go out in a pasture and probably have healthier feet, healthier legs. But when you have a free-stall setup, the cows are on concrete 24/7 – almost every situation today is like that.

Another important benefit to this type of barn is certainly in the handling of the manure used to fertilize the fields. As John notes, with a tank the size of the entire barn, the manure must be kept in liquid form for pumping:

You have to agitate it, otherwise you might end up getting all the fluid out first and end up with a whole bunch of solids towards the end. So when you're not pumping it into your spreader, which is hauling manure to the field, the manure is going through this pump, but gets pumped right back into the pit. That's called agitating it, but only when you're hauling it. When the tractor and the Husky are making a trip to the field, then the agitator is left on. The pump is left on to agitate. The pit can hold about four or five months. We usually manure the fields in June, and then when we get done planting corn, we empty out as much as we can possibly empty

out of it. Then in the fall, probably about the time we've got to start cutting corn, you've got to get more out, and of course it depends on how many cows you have. The more cows you have, the quicker it fills up. There's a little bit of barnyard-rainwater runoff that goes into it as well, which is probably a good thing, instead of it running off the barnyard into a water source somewhere, or the swamp. And you've got to have the water too. You'd be surprised – if you use a tremendous amount of bedding, it kind of dries things up. Been a few times we've had a hard time getting a pump to go down into the pit because it's solid, but after you get it in there and start agitating it, boy it loosens right up. That's the way to handle it though, because you can handle thousands of gallons a day.

While manure makes up most of the fertilizer used by the farm, they cannot avoid buying and using commercial fertilizers for the fields which are too far away for the tractors to get to economically. They may also decide to use chicken manure on a particular field. George notes the value and potential dangers of using this by-product on their fields:

Well, weed seeds may be in the feed that the chickens eat and it will pass right through them. I think it brought in velvet leaf, which we had here at one time, and it might have brought in this burr cucumber, although it pops up in certain places and it

Plate 133: Cold temperatures don't bother the cows, but they often slow progress on daily tasks.

Plate 134: George Palmer planting a corn-field with a six-row planter.

doesn't in other places. We've used it. We're spreading some this year; last year we spread none. If we do next year, it's going to be quite limited, because it drives your ph up too high, but you really get your money back in just the lime value in it.

After the manure is spread, the air surrounding the fields can remain pungent for a day or so, even after harrowing. George relates conversation overheard by travelers coming off the nearby interstate highway for a quick lunch break:

Either the Lewises had just spread something up here – they used to put some poultry manure up here, but they haven't now for five or six years – or we had put some of that stuff we've got down on

those two lots on the right, you know. A couple walks into the restaurant and one says to the other, "I hope the food's better than it smells!"

You know, you can spread it whenever. I tell them, 'Don't go down there – the wind's blowing right towards those places – wait 'till the wind shifts.' Or you do it right before it's going to rain, so hopefully it will dilute it more and wash some of it in.

As in any business, there are as many ways of doing things as there are people doing them, and farming is no exception. While there are many similarities in the way farmers in town pursue the tasks they tackle, no two do everything in a like manner. As John puts it so succinctly:

Everybody has different ways of doing things. That's why some of us use red tractors and some use green tractors.

This holds true for the planting of crops. Much of the land in North Stonington is very stony, and many years of picking these stones and stacking them into walls have gone into preparing the fields for crops. There are, however, a few sections in town where the fields are relatively stone-free, and plowing and harrowing don't bring more to the surface. In fields where farmers no longer want to pick up stones, they may employ no-till planters. In easier fields to work, they may harrow. The Palmer brothers, with cleaner soil than most, believe they get a better yield in their corn crop by harrowing. John discusses some of the advantages and disadvantages to each method:

Everybody has their own way of doing it. Some farmers don't do any plowing – they have too stony a land. Some farmers plant no-till – they don't plow or harrow. You just spray – kill the grass, kill the rye if you have a cover crop. Plant it, and come back and spray it again.

They have this corn now, different varieties of corn that won't die after you spray it if it's up. It's called Round-Up ready. It's a wonderful thing. You can go back in and spray your corn if you have some grass or weeds in different climates in the spring that could cause problems. The corn can be knee-high or so and you go back in and spray it and kill the grasses and weeds. Some farmers are really good at it. Planting no-till, they have real good results, and other times you see not-so-good results. One of the disadvantages is that sometimes the fertilizers don't get incorporated into the soil and the corn doesn't take up those nutrients. Nitrogen and potash will leach into the soil, but phosphorus seems to stay right on the surface. Personally, I think you get a little better results if you incorporate it.

We like to plow – turn the soil right upside down. Plow it, harrow it once or twice, and if you do a real good job plowing, sometimes you can get away harrowing it just once. When you're harrowing, you use disks – you're making the soil soft so you can plant it. Some of these corn planters nowadays, when you plant no-till, they have different set-ups on them. There are more springs to get the seed into the soil. I guess there's strong press-wheels on the planter so the seed is not exposed. You want to plant probably about an inch-and-a-half deep. We plow the Clark's Falls fields. In other places, some of the ones that aren't too stony, we might not plow every year, but every two or three years, and

Plate 136: Tim Blake brings a load of feed through the rain.

quite often get better crops. From our experience, you get better crops if you plow.

Some places in the country, they plow in the fall because the ground is too hard or too wet in the spring. They have to plow in the fall. They have these huge, heavy, heavy harrows nowadays that really do a job on the land. There are different harrows. There's a cutting harrow, and what they call a smoothing harrow. A smoothing harrow will make a smoother finish, especially if you are going to seed down with grass or hay. With corn it's not real critical that it's super smooth. But then you could put a drag behind your harrow, if you wanted to, and that will smooth things up. And if you have sod there it might help break up the sod – knock the dirt out of the sod. There's a lot of different ways of doing it.

Plate 135: Calf milk-supplement bag.

Crops play a major role in the life of each farm in town, and modern plant genetics have helped to increase crop yields, reduce labor, and minimize plant health problems for all of the area's farmers. John has seen the effects of this work, and sees other improvements to come:

There have been tremendous improvements – the genetics in corn. They are talking about corn now that will produce its own nitrogen so you don't have to add nitrogen to the soil – you don't have

Plate 137: John spreading solid manure onto a cornfield.

hay ground and put alfalfa in it, but it just hasn't happened.

As long as the weather cooperates, they begin planting their corn crop in early May. Once the fields are prepared by fertilizing, plowing, and harrowing if necessary, the corn is planted using a six-row planter. As John notes, additional elements can be put in with the corn at this time if desired:

You can put insecticide in a separate hopper that puts insecticide next to the kernels of corn in the ground. It helps keep the cutworms out, the root worms, but some of these varieties of corn now are resistant to that – the bugs don't eat them. You don't have to add that. And they have corn planters that you can put fertilizer into the ground the same time you plant. That's usually injected or run into the ground a few inches away from the kernels. It's not put right into the same ferrule, because it will burn the plant when it first starts to sprout, so it has to be put a few inches away. They have liquid and dry fertilizer for that.

Planting hundreds of acres takes time, especially if wet weather in the spring interrupts the process. As John remarked, this can work to their advantage. If they could plant all of their fields on a single day, it would have the effect of putting all of their crop eggs in a single basket. Having crops at different stages of growth may allow them to survive weather-related problems. The corn they plant is rated in days – days until it matures under normal conditions. While this can help corn mature at the time farmers would like it to, it doesn't always work out that way:

It takes us so long to plant,

Plate 138: Two corncribs.

to buy a fertilizer. That's going to happen some time. Insect-resistant corn, blight-resistant corn. It costs more money, because of all of the technology and time and research that went into it, but it's out there. Even ethanol. They have corn that supposedly will produce more ethanol.

Although seed companies focus most of their attention on corn crops, John notes that even the humble grasses have improved over the years, although he hasn't had enough time to take full advantage of these changes:

The grasses you don't have to plant every year. Some of the orchard grass and timothy and whatever comes in by itself and has been in the same sod for years and years. The alfalfa varieties are much better than they were years ago. Years ago, if you got three or four or five years out of an alfalfa stand you were doing pretty good, but they have better varieties now that last much longer. We have more than enough hay ground. It would be nice to have more good alfalfa. I suppose we could harrow up or plow up some of the poorer

if we planted it all the same day-season corn, it probably wouldn't make much difference, because you can't plant all of your corn in one day and you can't harvest it in one day. The corn that's planted, lets say, the second or third week of June, will come up a lot quicker, because the ground is warmer than for the corn that is planted the first of May. The corn that is planted early, you don't plant so deep in the ground either – it's more apt to rot. As the ground warms up and get drier, then you plant it a little bit deeper. The shortest season might be eighty-five or ninety, up to a hundred-and-twenty, hundred-and-twenty-two days. Usually the longer-day maturing corn, the bigger the corn is – you get more tonnage per acre.

While advances in technology can be helpful, they also keep farmers guessing what they should use. John continues:

You have to be a genius nowadays to pick out the corn because they have so many different choices. Insect-resistant, blight-resistant, Round-Up ready. Some of it's guess work. If you have real good luck with a particular variety the year before, or two years before, you might want to stick with that. There might be a new variety on the market that's got a lot of praise, and you might want to try that. With good corn, you get twenty tons an acre or better. We plant about two-hundred-and-fifty acres of corn or so. We have had over three hundred at one time or another, but then you lose some land.

When asked how many acres they have of the grasses they cut, John jokingly replies:

Plate 139: A retired workhorse.

Enough. We chop some hay, we chop some alfalfa. We bail some hay. Alfalfa is higher in protein. It makes a better feed. It's a good forage. Milking cows get a higher protein feed. If you can make some real good chopped alfalfa, you can save some money on your grain bill. It's high in protein. When you buy grain, usually the higher protein grain you buy, the more expensive it is. Usually the energy feeds are cheaper than the protein feeds. We buy cottonseed, brewer's grain, wheat mids. Some of the heifer grains have molasses in it to make a pellet. I guess that's quite important at the grain mill. To make a good pellet, they have to use some molasses to make it stick together. But if you buy just a grist, then it's not so necessary to have molasses in it. We have fed that before – just straight molasses – but it's kind of messy. I remember years ago, a long time ago, you'd put it on top of hay in the barn, so then if you had some poorer quality hay, it would be a good way to get the cows to eat it, but that's not such a good idea. Why put frosting on the cake if the cake's no good?

Like every farm, the Palmer Farm must import certain feeds for their herd throughout the year. John continues:

The grain bill is probably one of the biggest expenses we have, if not the biggest expense. Some grain comes every week, some comes every two weeks – it depends on what it is. Some might only come every six or eight weeks, like cottonseed. They bring a trailer-load of that, and it lasts quite a while. One particular grist is probably about twenty-two ton – comes about every two weeks. Brewer's grain comes about every six days. The heifer pellets – that depends. If you get three ton, it comes every week, and if it's six ton, it might be every other week or every three weeks – it depends on how many heifers we have on grain. You have a lot of money tied up in the animal before they put a gallon of milk on the shelf. It costs a lot of money to make a gallon of milk.

With as many animals as they have, George and John must hire on help

Plate 140:   Mark Taylor heads out to feed some of the road calves.

to keep ahead of the work a herd this size requires. As with the Lewis farm, they have several full-time and many part-time employees. The two largest farms in town also hire kids in the summer months to help with the extra work of handling the crops. John talks about the younger workers:

Plate 141:   John and Asa Palmer driving cows.

We've had some pretty good help over the years – even high-school boys. When you are on a farm you can hire help fourteen years old. A lot of fourteen-year olds can't get a job anywhere else. They can't work at a convenience store or a coffee shop until they're sixteen. With the high-school kids, they're pretty good. They usually get somebody else lined up to take their place if they're going to quit. It's not a snap decision, if they're going to go off to college or get a higher paying job, a cleaner job – whatever reason they might leave us. You take a young kid that doesn't have a car, no girlfriend – they don't need a whole lot of money. If they have twenty or thirty dollars a week in their pocket to buy junk food or something like that, they're happy and content. But after they get their driver's license and a girlfriend, that's when they've got to have more money… .

Mark Taylor is one of the full-time workers the brothers hire. Mark has worked for them since 1983, and he, too, comes from a farming background:

My family was in farming – third-generation farm. It started out originally as a poultry farm, back in the early nineteen-hundreds, and then we converted it over to dairy after the poultry industry went bottom-up – because so many people were getting into it – where you had to add on more chickens all the time to keep up with the times. It got to be too overwhelming for my grandparents, so they sold out, and then we took it over and started there as a 4-H project with just one cow and a calf. Then it turned into, we had forty-

five head of our own after a while, and show cattle as well. We ran a show cattle circuit for probably thirty-five years, along with the dairy farm. This farm is a mile north of the Palmer Farm, which is on Dennison Hill Road and Putker Road. We have about one-hundred-and-twenty acres of our own – it borders a brook, and we have three, actually four, active hay fields that we retail-sell hay now.

Although slightly different in age, Mark grew up with the Palmer brothers and enjoys both working with and for them:

I grew up with John and George. They're older than I am, but we're all kind of on the same page, so to speak. I started with them as a truck driver in the fall of 1983, in September, and it was going to be a two month bit. And they said, "When you're done, you're done." I've been here ever since then, and I'm not done yet. I'll be here probably until they close the doors. It's a relaxed atmosphere, as far as you can come and go as you want. If you go to the corner to grab a coffee, there isn't going to be somebody following you. I have Sundays off. I have two weeks paid vacation every year. I have about five or six sick days every year – paid as well – and full medical benefits.

While Mark may be found working in almost any area of the farm, his primary responsibility is with the younger cattle:

Plate 142: Tim pumps liquid manure from under the free-stall barn into the spreader for use.

My responsibilities are for the calves, from day one born, right up to eight-months old. I'm in charge probably of about three-hundred-and-seventy head out of the five-hundred head that they have on the farm. And that's my main job. Other than that, I do other chores. We might rebuild stone walls, we might work on fence, we might move animals around. I may drive truck still part-time. I don't milk anymore, since they put that parlor in. They were first breaking ground for that parlor back when I started in '83 and they milked in the old barn. I used to milk two nights a week, and when they got the parlor, they told me that they were going to have me go directly into the calf-feeding operation. That's dry cows, dried-off animals, and intermediate-sized calves, right down to the little tiny babies that are born day one. They call me the Calf Man. That's basically my job, but I am universal to everything on the farm when I'm needed.

As he is hired on as help, Mark's day usually starts at a more civil hour than does either George's or John's. Mark continues describing his typical day:

I'm probably there for 7:30 to discuss what we're going to do during the day. We kind of have a little business meeting. "OK, we're going to do this today, we're going to do this today. Don't feed these animals, because we're probably going to move this group around to another group." So we kind of dis-

cuss our daily routine, and then I'll start getting the truck ready – load it up with the feed for the very small babies. That's my first objective of the day, is to feed the very small babies, which are down here at what they consider the Gould property. I'm down there for about an hour and a half, and then I go back and I wash my utensils and dishes and whatnot. And from that time on, I usually start bagging-up pelleted grain for what we call the road calves, which are out on the road where I go with the feed wagon and a tractor.

I'm in charge of five different groups at this particular time. I load up my commodities on the feed wagon, as well as my grain, and then I go out to each group, one at a time during the day. We do a head check to make sure every head of cattle is in the group at the time, make sure there's water to them, do a quick fence-line check to make sure that the fences are in order and that nothing's going to be able to get out. A coyote could come along, anything can come along and chase them over the fence. Sometimes there's something out, and we get a call to go do that. We feed them at the same time every day if possible.

I'll start with the same group every day, and keep it in order as I go along, and check things as I go along. By that time, you are usually working up to the noontime, the one-o'clock hour, which I usually take an hour for lunch. Then I come back and I feed two more of my five groups, and then we usually kick around if we've got some other chores – cattle we may move around or work on a stone wall. We might go shuttle equipment around. This time of year, we usually shuttle equipment from field to field. That means you have to give people rides to and from the fields for the fieldwork, and then by that time, it's time to load the truck up and go feed the baby calves again. I then kind of wind down into the evening by doing my dishes once again, but I also check the cattle for calving before I leave every night. There's two groups of cattle that I check, and I make sure that everybody is all in check – that there's nobody having a calf or whatnot. Make sure everything's in order – the gates are shut where they're supposed to be and everything's on the right side of the fence.

While Mark might start each day at a more fashionable hour, it doesn't necessarily mean he ends it at one. During the crop season, everyone ends up working much later hours:

This time of year, we wind it down between six, six-thirty, but we go crop-work time, which is considered to be like chopping grass, haylage, and alfalfa hay-

Plate 143: A winter snowstorm covers the trees, fields, and a heifer.

Plate 144: Tim adjusts the knives in the corn chopper.

lage, when we could go as late as eight, eight-thirty at night.

The farm's cows, as the heart and soul of each of these operations, are carefully monitored throughout their life spans. They individually represent not only a significant amount of money, but a major investment in time and effort as well. For these reasons, the food they receive at every stage of their lives is carefully chosen. Mark details what each age group receives:

From day one, you can put a dish in front of them in which we have an eighteen-percent pelleted feed –

that's a special mix. No molasses. It's put in front of them from day one. They're kept on that pelleted feed along with good quality of hay and water, for two months, and they're also getting their milk formula at this time. After two months, they're taken off of the milk, and they continue to get the pelleted eighteen-percent feed, along with good-quality hay and plenty of fresh water.

After the two months they're moved into a group. We usually try to put them into groups of anywhere between ten and fifteen animals in a small, limited pasture. After about two weeks in that pasture, they're introduced to corn silage, which is top-dressed again with the eighteen-percent straight pelleted feed. When they're about eight months old, then they're strictly silage and hay and an astronomical amount of pasture, until the time they are brought into the barnyard, in which they're artificially bred. Then they will be fed straight silage and a very limited amount of a very stiff, first-cutting hay. They get their rumen working and whatnot, and they're bred at that time.

After that, when they become milk cows, they are on a very stringent diet of minerals, midlands – what was considered a brewer's grain – along with corn silage and chopped haylage, which is mixed thoroughly in a mixer wagon and fed anywhere between two to three times a day, depending upon the amount of animals that they're feeding at one time and milking at the time. They get that pelleted grain from day one, right until they're about eight months, and then it's taken away and it's basically just a silage mix. You may mix in a wet grain just to get a little bit of wet grain into them as well, but the pellets are actually taken away. Depending upon your pasture – if your pasture is desolate – you go heavier to haylage, because you want to get that roughage of the grasses into them. It also neutralizes the acidity in the corn silage. If you were to compare it to a person, they need the salad to go with the meal to help digestion and keep their systems working properly. Water is a very important factor. They need good, fresh, cold water all the time. They can dehydrate, and if their system shuts down because of dehydration,

Plate 145: Mark Rudd making welded repairs.

then you run into a lot of other serious problems.

Mark also keeps an eye on cows as they approach the birthing of their calves. While nine out of ten calves are brought into the world without intervention, Mark allows how there is always that tenth:

Let's say out of ten animals, you probably get one that has a problem. And a lot of time, and it's ironic, it will be in the night when no one is there, and that's when you really have issues. That's why it's important to check them the very last thing when I leave at night. George will come in and check them first thing in the morning. If they're having problems, or they look like they're going to be off, they will go off by themselves – you can kind of tell. If they're kind of hanging back by themselves, they're probably soon to calve. We can tell if there is a potential problem because of the way they get up and they step around. They have a certain step that they do. They kind of walk around, and do what we consider a waltz, and their tail is always back and forth, and they kind of act like they're trying to push and they're having difficulties. Usually they'll cramp up, and something isn't coming out just right. It may be something as simple as just one foot might be just cuffed back a little bit more than it should be, and you can reach in and just pull it forward quickly.

Plate 146: Corn just before harvesting.

Sometimes there's a tilt. Like the other day, I went up to an animal that I thought was having problems having a calf. It's supposed to be front feet first, with the nose resting on the hooves. I reached into this particular animal, and there was a nose and four feet – it was a set of twins, with one head and the front feet coming out first, and the back feet of the second calf. So when we pulled one calf out, the second calf did a complete clockwise turn, came right into position and we pulled that one out. But when I reached my hand in there, four feet, one head, I said, "We got problems." For that size farm, they're saying twins maybe every two or three years. We've had as many as six or seven sets of twins in one year, and two per month sometimes. We don't know why, and we've

Plate 147: Terry Oosterman trims the hoof of a cow.

also had triplets, which is very unusual. Five or six years ago, the farm invested in a defibrillator, which can actually bring back a calf that might have a slight heartbeat. We've actually revived a couple of calves.

As with any animal, cows have other health problems that must be addressed, but this generally doesn't take up much time throughout the course of the year. John discusses a few of the more common problems that have arisen:

They have problems. They can get pneumonia, a fresh cow can get ketosis, or milk fever. It's a big thing when they give birth. They lose a lot of water, they lose a lot of weight. If they have a difficult birthing or they have a pinched nerve – a lot of things. We have a vet that comes in that does pregnancy checks and vaccinates calves. If we have a sick cow, we don't always know what the problem is, especially if they have a displaced abomasum, the vet will work on them and try to correct that problem. You can vaccinate them for respiratory problems, to help cut down on mastitis – coliform mastitis, when they're in their dry period.

Plate 148: John scrapes the manure from the barn and the yard into a spreader.

While George and John may be at the farm seven days a week, Mark gets Sundays off to relax with his family:

Sunday comes, and then "Thank-God for Sunday" – and I say that almost every Sunday. Not that I don't enjoy my job, but I love my Sundays. I relax, I spend time with my dog and my family. We have a big Sunday dinner – a sit-down Sunday dinner that we've done since way back in the early days. We always eat anywhere between two and five, depending on if we have company come over, which was a traditional thing that my grandmother set up years ago. If we had hired hands that were out helping with the farm, they were also welcome to come in, grab a plate and sit with us and eat during the evening, and converse and whatnot. We've done that ever since I can remember; her grandmother did that. It's just something that's been carried on, and we still do it.

As with many farms, the Palmers keep a bull to impregnate cows not bred through artificial insemination, but as with all of the dairy farms in town, AI assumes the major role. When asked about the role genetics has played with their herd, George is perhaps less enthusiastic about its benefits to them in particular:

I don't know about the science end of it. Artificial breeding probably has been a big thing, but maybe we're right here not getting the results that we should. But there are a lot of other management factors, too. Some people are just better at getting milk out of a cow than others – that's all there is to it.

However, they do take advantage of AI and whatever benefits it brings with it. As John says, they hire out this particular work:

With the artificial breeding, it's up to the technician. He belongs to an AI company. We hire somebody that comes and breeds the cows artificially, and it's more or less up to them. If a particular animal has a bad trait, we might suggest to the technician what that animal needs to be improved on – bad feet, bad legs, or slow milker. These bulls that they use,

Plate 149: Neil Main sprays manure onto a field before harrowing and planting.

tremendous records have been kept on their offspring to tell all the farmers what particular traits the bull offspring will help you with. We have another guy that comes around once in a while and he walks through the herd, and he looks at each particular animal like a judge. He walks through the herd, and he sees a particular cow that has real wide teats on her udder. Well, you want to breed that particular animal to a bull that will supposedly correct that bad trait, and the next generation the teats will be closer. Of course, this fellow that checks the herd doesn't know if it's a slow milker or a fast milker, and that's important in a milking parlor. You don't want to have too many slow milkers, because you've got eight, nine cows on a side, or how-ever many cows you've got in a parlor on each side, and with one slow milker the rest of them have to stand there until she's done. If you had a stanchion barn, where you milk each cow individually, well, you carry the machine to the cow; that's

not such a big problem. It's not tying up so many. So you don't want to have too many slow milkers. I guess if you had the perfect setup, you could put your slow milkers in a certain area of the barn and milk them all together. That would be pretty good. When you spend that much time milking them, you know which ones milk out slow, which ones kick, and which ones are miserable to milk with.

As is often the case, there is no one clear-cut answer to the question of what is the useful life span of a dairy cow. Every farmer asked wished genetics would increase the working lives of their animals, and this is no surprise. With two years of feeding and care before any cow begins to make her first gallon of milk, the importance of keeping them in the herd and productive after that period is obvious. Each farmer, however, uses different criteria to determine if a cow will remain in the herd. John remarks on some of the factors in deciding which animals to keep:

If the cow is making money for you, you keep on milking them, regardless of how old they are. Maybe after five or six lactations they might start to go downhill and not get so much per lactation. Usually the first lactation is not so much, and then the second, third, fourth, and fifth are much better. It probably varies from animal to animal, which lactation they peak at. That can vary from animal to animal, because they might have twins. If they have twins, that takes a big toll on them. Usually the body condition is not so good when they calve or a cow that has twins versus a cow that has just one. We have control boxes on our milking equipment so you know what their production is, and if you milk them often enough, you know which ones give milk and which ones don't – which ones are not earning their keep. If they don't earn their keep, they go for a ride. I'm sure some farmers have a cutoff limit. If a cow is not bred, and if she's giving less than thirty pounds a day, she might go for a ride in one herd, and it might be forty-five or fifty pounds in another herd. It's hard tellin'. You know Tricia's cows give a lot of milk, and I'm sure she sells cows that we'd hold on to. Other farms, if they give twenty or twenty-five pounds, they might hold on to them. Who knows?

One distinctive, and perhaps essential trait common to farmers in town is their independent nature. One cannot help but feel that a large reason they farm is because they love the freedom their work gives them. They make, for better or worse, all of their own decisions based on what they feel is right. As John says, this can run from decisions on what breed of animals to have in the herd to which grain companies should supply their feed:

We all can't agree on the same thing. It's just like buying grain. There are different grain companies out there, and who you buy grain from is different from who I buy grain from. It's just a way of being independent – it's the way it is. There are more grain-company choices than there are milk-company choices, that's for sure. We get some from Ventura; some comes from Feed Commodities; the brewer's grain comes out of Newark, New Jersey. I don't know where the cottonseed comes from down south. South Carolina, North Carolina? George makes the connections on that – he deals with a broker, I believe.

Plate 151: John drops some hay into an outside feed rack.

One option farmers have little choice with today is where their milk will go. At one point in time they had many choices, but with fewer and fewer farms surviving in Connecticut, there are fewer and fewer processing plants able to handle their product. While the Palmer Farm's milk is picked up by Agri-Mark, John states that he still doesn't necessarily know where it will go:

Plate 150: Tim chopping corn.

Agri-Mark milk probably goes to Garelic, too. Trucking is a big thing. They don't want to truck it any farther than they have to. It might depend some on supply or milk in the area. If there's too much milk going to Garelick, then some of it might go to Springfield, where they can make powder out of it. Agri-Mark has a plant up there in Springfield, Mass. I believe they can make cheese. Some processing plants can make powdered milk. As far as I know, Garelic

only deals with fluid milk. But they package it or bottle it for more than themselves, so it might say Shaw's, or it might say WalMart or Stop & Shop. You just have to look at the plant number on the container. That will tell you where your milk came from. Shaw's doesn't have any dairies, and Stop & Shop doesn't have any dairies, so everybody's milk goes into a big silo and it gets pasteurized and bottled up.

While New England dairy farmers face a host of problems, from loss of land to increased costs for grains and fuels, the overwhelming root of their troubles can be ascribed to the low price they receive for raw milk. In the year I spent working on this book, farmers in town were paid, at a low point, ten dollars for a hundred pounds of milk, or approximately $1.16 a gallon, whereas the cost to the consumer off the shelf was nearly four dollars. A running joke in the farming community is that there are only two people who understand milk pricing – they're both in Washington and even they don't agree with each other. Reactions to this question amongst local farms is nearly universal, and John's answer is typical:

Plate 153: The homestead in winter.

You've got to be an Einstein to figure out how they price the milk. It's supply and demand, how much cheese is in storage; how much butter is in storage. You can't keep butter in storage for too long because it will go rancid, but cheese can keep for a long time, and why it fluctuates, I don't know. The store – the retailer – makes more per gallon than anybody else. And of course the trucker has to make some, and the processor has to make some, and if there is anything left, I guess the farmer gets it.

George goes on to discuss the intricacies of the price of milk in relation to both cheese and butter, and notes that even with the slight rise in milk prices recently, other factors remain beyond their control:

This whole milk-pricing scheme is very, very complicated. You'd have to get a doctorate and even then I don't know if you'd understand it. Really it's based on the milk that goes into cheese and butter more than fluid milk. It's based on the marginal product. There's four classes of milk. Class One is fluid. Class Two depends on what products there are, but Class Three is really what it's based on, which is cheese and butter. Powdered and whey is Class Four. There's been times when they'd even charge you for some of that. Deduct you because they had to process it – low demand. Things are supposed to get better. I don't know – they've said that about the whole economy. I guess milk prices are better, but inputs haven't dropped very much. Grain prices were a little bit lower, fertilizer prices were a little bit lower. Some things never go down – you can't really beat your labor costs down. You cer-

Plate 152: George driving one of the farm's many John Deere tractors.

tainly can't beat your insurance costs down. You can't beat your local taxes down, although we have some tax breaks that I never thought we'd have.

Of course milk pricing is going to fluctuate as availability goes up and down, just like any other product on the market. While there are few dairy farms left in Connecticut – just around one hundred and fifty – this doesn't necessarily mean scarcity in the marketplace. Whereas all milk used to be local at one point, as George points out, this is no longer the case:

And that's another thing – the transportation system. Even with tougher restrictions on driving time, they can shuffle milk around pretty fast. In the summer-

Plate 154: Neil Main and Tim Blake confer while chopping corn at a field distant from the farm.

time there's a lot of milk that moves to Florida because you can't make milk in the heat. They're always talking about how many truckloads went to Florida in the summer and how much excess they have during the wintertime. I don't know if the milk production in eastern New York is what it used to be, but you can get it in from central New York in no time at all on the thruway. Three hours and it's in New Britain, you know? That Route 17 in southern New York – they used to call that the Milk Turnpike because it was built to get the milk off the southern tier into New York City. That's how much pull the rural legislators had up in Albany. There's not much else out there.

But beyond the trucking of milk from distant farms, there are other factors

creating problems for the town's dairies. Farming is, as George points out, no different from any other business where a product is created:

I've always said, no matter what the product is, it's a case of killing the goose that lays the golden egg through overproduction. It goes in cycles. The big thing right now that's causing a lot of this overproduction is sexed semen. You can buy semen that's ninety-to ninety-five percent certain that your calves will be females, you know? I didn't think it would catch on like it did, but apparently I was wrong, because there's a lot of heifers, and when there's a lot of heifers, it means milk production won't go down. Even in this crisis, you know, this tough period we're going through right now, it might last a long time because of that and because the people that are financing all these big operations don't want to pull the plug on them. They're not going to get their money back, so they're stringing them along, and there has been a fantastic loss of equity.

And, of course, farmers are no different from the rest of us in what they face in the economy or how they react to it on their farms. As George notes:

There's also this thing of keeping up with the Joneses, you know. I guess that's endemic with a lot of farming. "This guy's going to beef up to two-hundred – I'm going to milk two-hundred-and-fifty, and I'm going to be big-

ger than him!" If it's all organic growth – something you can generate in your own business one way or another – say the potential was always there but never utilized, or you bought another farm, or rented some more land – it's a little more risk, because you can build facilities and then find you've overbuilt. It's a challenge – it's no different than a lot of other businesses. Construction people – same thing. Things will be good, they're making all kinds of money, they're buying new equipment, putting on more people, and all of a sudden it disappears.

His brother echoes some of the same feelings on farm expansion:

Production has improved, but I don't know that we are any better off today than the farmers were fifty, sixty years ago. I really don't know. If you get more family members involved, you've got to make more milk so everybody can make a good wage. If you have the land that will grow enough feed for a couple of hundred cows, why not do it? If you get good hired help, you should make money off your hired help. If your hired help is costing you money, why bother? Equity – the more cows you have, your net-worth is more as long as you don't have to borrow money – that's not the way to go.

The trend throughout the United States over the past twenty years has been toward centralization into fewer and fewer large corporations. Three companies now control nearly ninety percent of the milk produced in the United States. As George points out:

Plate 156: A beautiful stream runs through the Palmer Farm.

Plate 155: John and Chris gather straw.

You know the whole tax system is set up for growth, but do you get to the point where you just have one big entity that controls so much? A perfect example is WalMart, you know. They claim that their thing was that they've always had fantastic control over their suppliers. They say up at the milk plant if the federal inspector is there, or the state inspector is there, they don't get too excited. They do get excited when the WalMart inspector comes through, because that's their market. It almost gets to the point where the tail is wagging the dog.

Dairy farming has also moved from being a family business, passed on through gen-erations, to something looked upon as a quick investment strategy by financiers wishing to make a buck in a different field:

Well, they say out West, there will be nothing there, and then the next year there will be a five-thousand-cow dairy there. It's all financed. I'm sure as hell if I had that kind of money, I wouldn't go out and build a five-thousand-cow dairy. The money has been there to lend, the technology is there – that's another thing – and cheap labor, whether it's legal or illegal – that's made things grow out in California for years. Victim of, I don't know if it's success? It's a lot of different factors. A lot of times, and it doesn't matter what the business is, it's expansion, because the lenders know they're not going to get their money back, so it's a way of hiding it – the debt. They say, "Well, you can't pay for it with two hundred cows, but maybe you can with four, so we'll expand and we'll just roll that past debt into it. We'll start all over again." I'm sure that's the basis for a lot of these expansions – no matter what it is – any kind of entity. I guess it catches up sometimes. I've mentioned that thought to people, and a lot of them agree with me.

Beyond changes in market forces or the progress made in technologies employed by farmers, George sees the subtle changes in society affecting dairy farmers as well. He feels the way in which society now looks at farming is much different than it used to be:

It's an awful big change. Years ago, even if there might not have been any more cows in town, there were certainly more farms. A lot of people had worked on a farm at some point in their life. Maybe it was some real menial work, you know, because there used to be a lot more shoveling this and shoveling that. You didn't have a bucket loader do this, or you didn't have barn cleaners. Bend over to milk cows. Cutting corn, you didn't have these monstrous machines – you'd cut it by hand and throw it in the truck. You'd run it through a stationery corn chopper and stuff like that. A lot of kids worked on the farm a little bit when they were growing up – "I worked for your grandfather – I got fifty cents and hour!" Well, you were happy to get the work, and who was else was going to pay any more anyway? It's not like anybody got rich on it. These people in town today, number one, even if they grew up in town, most of them haven't got the slightest idea about it, and then the ones that move in and out, they know even less. They look at you like you're some kind of a zombie. I guess

Plate 157: Neil Main and his wife, Christina, delivering chopped corn to the silage pit for storage.

they call it a cultural difference. Some of them understand – there's a few that understand. It's not just here, it's the whole country. People don't understand where their food comes from – period – or what it even is they're even eating. No idea.

As farm after farm has disappeared, another essential element went with it – the sense of community amongst dairymen. George remembers what it was like earlier in his career:

The big changes – there's cultural change. Like I said, other people are completely ignorant of what's going on, and you also don't have anywhere near as much of a peer group, you know, where you can go talk shop. That's an important thing – it really is, because it keeps people motivated. They had that commission sale up there in North Franklin. Christ, when I was a kid, and even into the early eighties, that was a social event too. They'd take cattle up there – beef and heifers. Jim Kahn had a big barn there, longer than this. They'd fill it up with dairy cows every week. People would swap cows every week – two or three. Bring a beef cow in, go home with another cow. They had a little arena there where they sold them, and the place would be chock full. Some people went every week, no matter what they were doing. You could have fifteen acres of hay out and you still had to go up to the auction for a few minutes, you know? The fellow that visits Dan over here – George Whitford – he lives up in Rhode Island. He always went there, every week. That peer-group thing. It's no different than anything else. That was all pretty neat. In England, when these guys get done at the end of the day, they just head to the pub, have a few drinks, and talk over the activities for the day. Then they go home and start all over again the next day, but you don't see that so much around here.

Towns in Connecticut and neighboring states have made some effort to keep farms in their communities. Given the vast amount of land it takes to raise feed for their animals, tax relief is generally the form in which this assistance arrives. George:

We've had this hundred-thousand dollar exemption on equipment – farmers and fishermen are both qualified for it – but they can spend a hundred-thousand dollars on one piece of equipment very fast, right? Well, the same thing on a farm. You can buy a tractor and pay a hundred-thousand dollars for it, so one piece of equipment can gobble that right up. This town they did it so the first hundred-thousand on your buildings, other than your house, is tax-exempt. The whole thing is, what qualifies you as

Plate 158: This bog, once used to grow cranberries, now serves as a resting place for waterfowl.

Plate 159: Working cows inside the free-stall barn.

a farmer, you know? That's one of the cloudy areas in some of this. Some assessors are real hard about land use and personal property like that. Others are a little more lenient. I think it depends on the whole philosophy of the town – what they want to see stay there. I'm sure there's people that go by here – we just spread manure – and people probably don't like it. Some people can live with it for a short term and realize they're better off with it just like it is.

Although Connecticut has made some effort to protect and save farmland from the pressures of development, farmers in town have yet to take advantage of the program. George notes that Rhode Island has done a better job with this:

Of course you have this development-rites scheme in Connecticut. It's good and it's bad. It's good because you know that land will never be built on, but it takes the value right out of it if you own it. A lot of times, a lot of farms in the state, they sold the development rites to the state and the next thing they do is sell the cows! I've seen that a number of times, you know? But the farm is still open – someone else will use it or they'll raise some hay or something else. They've relaxed standards on it. The whole problem with Connecticut is they never really put a lot of money into the budget. It's funny, in Rhode Island they seem to have more money available for it because I think there are some foundations and trusts that kick in a big chunk or buy it outright. There's a Champlin Foundation in Providence – they must have been big mill people or something – but they've helped buy an awful lot of cropland in Rhode Island. They really don't have that in Connecticut, and it's made a big difference.

Keeping farmland open, however, may have little to do with the survival of dairy farms in New England. There are pressures and realities that can overwhelm even the most ardent farmer's desires to stay in business. John believes there will always be the need for what he does, with some reservations:

I think there will probably always be a market for milk. It's really a terrible business. Dairy farming is seven days a week – it's very demanding. Somebody has to be there to milk the cows, to feed them, to clean up after them. Milk truck backs up to the door, takes your milk, and you get paid what they want to give you – you can't set your own price. You have a culled cow, you send her to the auction, you can't set your price. They'll send you a check for what they want to send you, I guess. It's a terrible business – we can't set our own price. But we go to buy something and we can't dicker and get them down on feed, or fuel, or supplies. It's hard to get anybody to come down. We can't set our price – their price is already set. The profit margin isn't very high. And it's too bad that you have to

Plate 160: Tim Blake dumps a load of chopped corn.

have these give-away programs to keep us in business, because it's not the way it should be. Even the 490 Act, where the land gets a tax break in town. In one sense it's good. We don't have to pay building-lot prices on an acre of land, but on the other hand, why can't we get enough money for our product so we can pay taxes like everybody else does? The average resident wants to have scenic views and likes the farmland. It's a crazy business. There are a lot easier things to do in life than milking cows.

George is perhaps more troubled about the future of dairy-farming in the area. As he says about the total number of farms operating:

It's always gone down, so I don't know that's ever going to change. I don't know how low a figure it can go to. There are places in this country where there are no dairy farms period. Long Island doesn't have any anymore. The last one sold out probably twenty years ago. No more dairy farms out there. The numbers have gone down. The peak production for pounds of milk in

Connecticut, I think, was about 1991. The thing in Connecticut is the fantastic pressure on land prices.

He also knows that farms are not disappearing only because of the value of the land they occupy. Times are much different from when his father farmed:

Plate 162: Tim harrowing a cornfield in the spring.

It's not just profit. No one might be interested, or there might not even be a family to be interested – they might not have any children. You know it can only go so long, or the kids that you have aren't interested. It could be financial – it could be estate-tax burdens. It can be too much debt load over the years, so finally they have to get out of it. Divorces – no different than a lot of other businesses. There's a lot of other factors besides just the bottom line. Some just say, "Hey, go do something a little easier in life." They don't even want their children to do it. Sometimes they'll piss and moan so much that why should the kid be interested?

Despite the ups and downs, the farms in town continue on with what they have always done and hope for a better future. George notes the uncertainty in the business today:

It's day-by-day right now. It's not a really good environment right now. They thought prices were going to go up more than they did last fall, and then they've come back down some. Now they're predicting there is going to be more of a global demand. There's some areas in the world where they could do a lot in the dairy business and they don't. The other driving factor – as poverty decreases in the world, those people will have more of a demand for protein products – meats, cheeses, and stuff like that. We'll see. You know, you've got a billion people in China. If the per capita was even ten percent of what the consumption is in this country, that's a fantastic demand.

Plate 161: George stops the planter to check the seed level in the hoppers.

Plate 163: The cows wait expectantly in the barnyard for their food to arrive by tractor.

It is not for everyone. You have to be married to it in a way. Your availability is one-hundred-and-ten percent. You have to be able to stick through the long nights and the early mornings. It's all what you make it. It's like anything that you're doing. If you are happy with what you're doing, it's not considered a big thing. To me, it's just another day. I'm here, I'm alive, I'm on the green side of the grass, and I love it. You can't beat it.

Mark is also worried about the future in farming, and is happy he is not in the position held by his employers:

I feel like, right now, I would not want to be owning a farm of my own. I would rather be working for people. I've said that a hundred times over, especially in the last year or two. It's a bleak outlook on things – it's not good at all. It's just a matter of them being established for so long and having their heart in it so much, that's keeping things pretty much going right now.

Despite those misgivings, he doesn't see anything else he would rather do:

Plate 164: John and George in the milking parlor.

**Scenes from the
Palmer Farm**

Valley View Farm

About the time I was wrapping up my work on the last four dairy farms in town, a new farm sprung to life. Ed and Belinda Learned purchased a barn and its adjacent field from the Bill family to add a dairy component to their existing operations. The work on the farm will be managed by their three sons: Ben, Tim, and Will. Tim talks about their family's early history in farming:

Over the past twenty years, our family started raising three beef-steers over at our grandparents – we used to get silage from the Bills and the Lewises in five gallon buckets, and over the years our herd has slowly increased in size and we got more and more involved. We created larger and larger pastures and now we have our own herd. We also used to raise replacement milking-heifers for the Bills. My parents have always had a big garden and the focus was always towards making better produce for the family, and now we do it commercially.

Besides tending to their own animals at home, all three brothers worked on the Palmer or Lewis farms as they got old enough. Tim continues:

Ever since I was about thirteen years old, I worked

Plate 165: The barn and milking parlor at the Learned's new Valley View Farm, LLC.

on the Palmer Farm, and George and John taught us a lot about dairy farming. Now we have our own barns – it's sort of the culmination of the experience. When I was really little I was just haying and I was being taught how to milk cows in the barn. I got paid for haying but never for milking, and then when I turned thirteen, George gave me a time card, and then I graduated from milking to crop work and tractor operating. I always had a hand in the milking parlor and other barnyard chores – mostly cow work.

Ben, the oldest son, also worked for the Palmers:

Growing up I worked on the Palmer Farm, right down here on the corners, for probably five years – from probably fourteen to nineteen years old I'd say. Started out probably doing simple things, putting in fence posts, fixing fence, acting as labor to help bale hay. Over time I got to do more and bigger things.

Will worked there as well, and added this about the start of their new venture:

Plate 166: Tim and his mother Belinda at a farmer's market.

This farm went up for sale, and my dad asked us, "Do you guys want to farm – dairy farm – and really go for it?" And we decided, "Sure! Why not?" It's nice working for yourself – knowing you are reinvesting your own money – you are not putting in all of your time for somebody else's business.

Tim talks about his father's interest in dairy farming as the catalyst:

My dad has always wanted to milk cows. He had an uncle who did it, and he liked spending time at his uncle's farm over the years when he was a little kid. Then his uncle sold the farm so he decided he would have to start his own. So we started up in the woods and now we are slowly working our way into the valley. Eddie Bill retired from the dairy business quite a few years ago, and then his parents decided to sell the land because they had no more interest in it, so we bought it. We put in corn for our first three years, and then our fourth year we finally started milking. We started out buying heifer-calves. When they were a couple of days old we would buy them, then, when they got older, we bought a semen tank and bred the heifers we had. Then it finally came time after we signed our DFA contract – the barn had power and everything seemed to be lined up. I think at the time we were milking five cows into a can. We were just using the vacuum system to run the stainless steel surge can, then we'd feed the milk to the pigs. Finally we thought we had enough cows to fill the tank enough to finally hit the agitator so the milk wouldn't freeze on the bottom and we called up Tomaquaq Milk Hauling and set up our appointment. I believe it was August 1st – that would be our first milk pickup – and they have been coming every other day ever since. When we started out it was about 400 pounds we would produce, and now we are up to producing about a ton of milk every two days.

Starting a dairy operation at this particular point in time is certainly a risky move. Ben notes:

Plate 167: Without a pit to stand in, Tim must crouch to attach the milking machines.

It was interesting. The price of milk was very down when we first started – that helped us out in some ways, not in others. It made acquiring dairy cattle more affordable, but the price of milk barely covered bills as it was. I think the thing that stops most people from starting up dairy farming is the investment. It's a huge, up-front, capital investment for marginal profit. I guess you just have to milk as many cows as you can and improve your margin – that's our goal.

As with any new business, there were bound to be good and bad days. Tim, like his brothers, feels that on the whole they are doing well:

It is a very interesting way to make a living. When we started we thought, "This isn't so bad." Now it's getting pretty bad – we don't get much sleep. We are pretty ambitious, and ever since we started, these are the first two weeks we've worked backward in fourteen months, and it's only because we switched grain. A couple of minor setbacks – this is the first half-a-month with negative performance. Starting a dairy farm with all of your investment capital in a period of the worst milk prices on record is kind of a risky endeavor. I'd say it worked, but we'll have to see how long the price remains in the positive section. So I'm optimistic – I think we should to be able to make it.

Plate 168: Ben spreads a ground-cover of wheat on the cornfield for the winter.

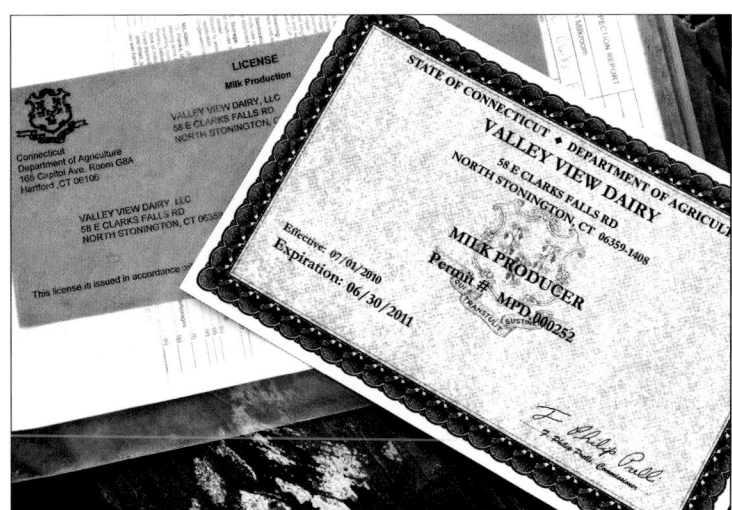

Plate 169: The farm's license allows them to begin milk production.

The Learneds also benefit from a close-knit family structure, as Will points out:

Me and my two brothers, we do pretty good. It's surprising, once you get everybody together and you all push in one direction, it's surprising what you can get done. It's nice with my two brothers – it's really nice.

As on any farm, each person ends up doing a little of everything. Ben describes the division of labor he and his brothers have settled upon:

We've kind of established over a period of time, by trial and error, who does what. We all have our different jobs and on different days we'll fill in for each other, to get some time off. It's not really a set thing. Willie does most of the chores up at the house; he feeds the chickens, the turkeys, the pigs, the heifers, the heifer babies. We all kind of do the beef at the house – who ever notices that they need it – and either me, Timothy, or Wade will feed these beefing heifers down here. Recently it has been changing because I've been getting my brothers on the tractor more. Since I was probably thirteen, I was always the one to mow it, ted it, rake it, run the baler – probably because I was the oldest – and that kind of became the way it was. I didn't really like that,

Now, having gone full circle around the barn and the fields, we find in the end a new beginning. Tim mentioned in passing, that on the day his family applied for their dairy license, two other farms in neighboring towns had applied as well. Perhaps there is hope then, that as long as there are young men and women who dream of their own farms, the family owned and operated dairy will continue to exist.

Plate 170: Ben and Tim load wheat into the spreader from a two-thousand pound bag of seed.

because it put me at a disadvantage time-wise. It's good for my brothers to get out there and operate equipment also. The last two or three years we've picked up a lot of bigger fields where they are not so technical as the little tiny fields we used to do, where there are rocks and things to hit or avoid. In the past two years my brothers have both become pretty good operators.

And so the Valley View Dairy is off and running. The brothers all share the same optimism, and as Ben comments, they intend to spread their chances for success:

I think you have to spread out and not just milk cows. I think you have to sell forage if possible. We also raise beef cows, chickens, meat chickens, turkeys, pigs – we pretty much do it all and I think that helps out a little bit. I don't really know how it's worked out for us so far – it just has. It's kind of like anything else – you get out of it what you put into it. If you don't let your life revolve around farming, but just let farming be your life, it seems to work out.

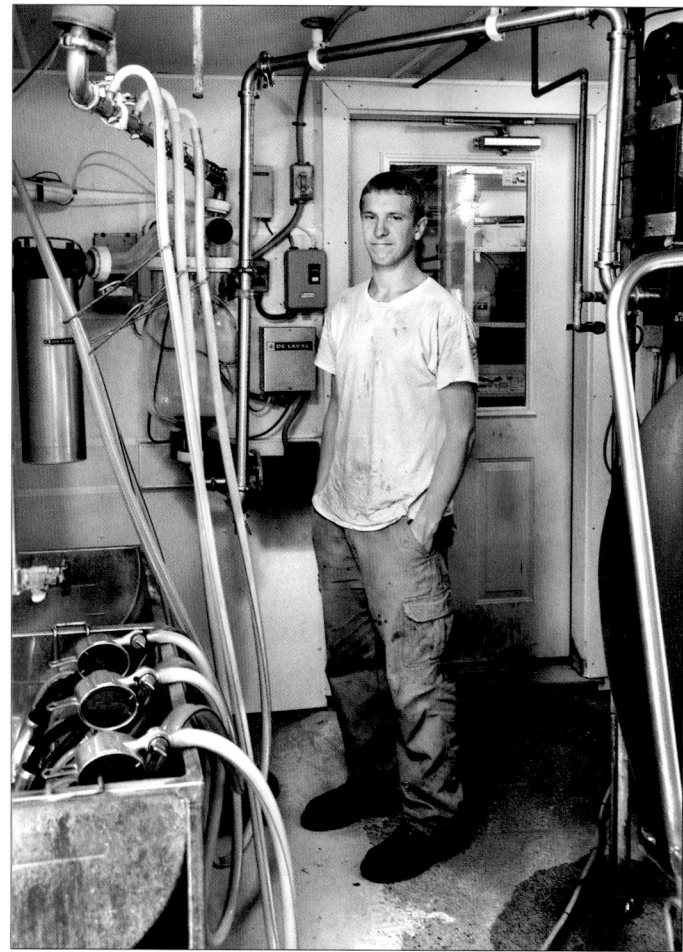

Plate 171: Will Learned in the tank and wash room.

"I had rather be on my farm than be emperor of the world."

- George Washington

Acknowledgements

First and foremost, I would like to thank all who appear in this book. Despite the enormous amount of work that every farmer must accomplish daily, each was more than welcoming during my frequent intrusions into their lives. Granting me complete access to their lands and buildings, as well as the freedom to photograph whatever I chose, they also tolerated me chasing them around their fields with my cameras as they tried to keep ahead of their work. Despite the rare moments they have to relax during the day, they were also generous in their willingness to answer my questions and shed light on the intricacies of their operations. Without such an open welcome, these photographs would not have been possible, and I thank each and every one of you for all of the help you have given me on this project. May your farms and families continue to grow and prosper.

I would like to thank my oldest sister Debbie for taking the time to edit this piece, eliminating what seemed like several thousand grammatical and punctuation errors of my own making. While not even the rack could pull my writing into the shape of something beyond merely tolerable, she at least made the valiant effort to have it conform a little more closely to the rules and regulations of the English language, and for that I truly thank her.

I would like to thank the North Stonington Historical Society for so generously deciding to publish this volume. Despite the many pressures on their budget, their belief in the importance of the story of the town's farms made the printing of this book possible, fulfilling their mission as safeguards of local history.

On a personal level, I would like to thank my wife Sue. Why and how she puts up with me as I pursue these projects is well beyond my power of comprehension, but I am forever grateful that she continues to do so. Forced to look at thousands of photos and read endless amounts of text at the end of her own busy day, she never complains, but rather, offers new encouragement and insight. Beyond being the love of my life, I could travel the world over and never find a better friend.

And finally, I assume all responsibility for any errors that appear in this work. Despite each farmer's effort to educate me about the dairy business, mistakes may still have crept into the text. Please be assured, they are all of my own making.

Markham Starr